CONCEPTS OF FORCE

A STUDY IN
THE FOUNDATIONS
OF DYNAMICS

CONCEPTS OF FORCE

A STUDY IN THE FOUNDATIONS OF DYNAMICS

Max Jammer

HARPER TORCHBOOKS · THE SCIENCE LIBRARY
HARPER AND BROTHERS, NEW YORK

CONCEPTS OF FORCE

This book was originally published in 1957 by Harvard University Press and is reprinted by arrangement

First HARPER TORCHBOOK *edition published 1962*

Library of Congress Catalog Card Number: 57-7610

Printed in the United States of America

PREFACE TO THE TORCHBOOK EDITION

The present republication of the book gives me a welcome opportunity to correct some errors of the original edition. It also enables me to express my appreciation to all those whose valuable criticisms stimulated further work on the subject of this study.

I acknowledge my indebtedness in particular to Professor E. J. Dijksterhuis of the Universities of Leiden and Utrecht and to Professor R. B. Lindsay of Brown University for having drawn my attention to a number of inaccuracies and omissions in the original text. I am also grateful to Professor Jules Vuillemin of the University of Clermont-Ferrand for his comments in the *Annales d'Astrophysique* (1959) in which he rightly takes issue with my omission of Martial Gueroult's important discussions (published in the *Revue de Metaphysique et de Morale,* 1954) on the metaphysics and physics of force in the systems of Descartes and Malebranche. Gueroult's—and I may add, Vuillemin's own—investigations clearly demonstrate the important role of metaphysics in the development of the concept of force from Descartes to Kant.

I am also indebted to Dr. Mary B. Hesse of Cambridge (England) for her competent comments in *The British Journal for the Philosophy of Science* (1959) and to Professor G. K. Konyk for his penetrating analysis of the book in the *Voprosi Filosofii* (1960); their objections, though on different grounds, to the philosophical tenet of the present treatment certainly deserve a more adequate discussion than a preface could allow.

Bar-Ilan University, Israel
September 1961

M. J.

PREFACE

In our present age of rapid technological progress the frightening discrepancy between our technical "know-how" and our philosophical incomprehension, in general, of basic scientific conceptions seriously endangers the integrity of our intellectual outlook. The cogitative activity of the modern scientist, who is more a technician than a philosopher, is strained to its utmost limits by the necessity for digesting the swiftly accumulating information in his specific field of research. He has little opportunity to indulge in the fundamental problems relating to the very concepts which he applies. Moreover, in our present system of academic instruction a thorough and critical discussion of fundamental and apparently simple concepts in science is consciously omitted (relegating them to a stage where the student's mind is still too immature to understand their true meaning). The nuclear physicist, for example, who works on exchange forces (of the Majorana, Bartlett, or Heisenberg type) and who discusses noncentral tensor forces, rarely has ever analyzed the concept of force in general, a concept that is absolutely fundamental to all his work. A historico-critical analysis of the basic conceptions in science is therefore of paramount importance, not merely for the professional philosopher or historian of science.

The present monograph deals with the history and significance of the concept of force in physics. Although placed in the front rank of our scientific conceptual scheme, the concept of force has never before been subjected to a comprehensive historico-critical analysis. Even today it is not yet generally recognized that during its long history the concept of force developed into an auxiliary or intermediate concept of primarily methodological importance. Only too often (and not only in elementary or introductory courses in physics) "force" is interpreted in the traditional animistic or Peripatetic manner as a "tendency" or "striving." Only too often is it stated that one force "overcomes" another force, an interpretation not unlike St. Thomas Aquinas' comment[1] that the velocity of a moving object has its origin in the "victory" of the moving force over the mobile or its resistance.

It is the purpose of this book to clarify the role of the concept of force in present-day physics. It is pointed out how ancient thought, with its animistic and spiritual interpretations of physical reality, laid the foundations for the development of the concept and how in preclassical science the concept of force became invested with a multitude of extrascientific connotations that greatly influenced the interpretation of the concept until very recent times. In fact, their effect is still easily recognizable. It is explained how Kepler initiated the scientific conceptualization of our notion, how Newton groped for a clear and profound conception and how post-Newtonian physics reinterpreted the idea. The concepts of Leibniz, Boscovich, and Kant are confronted with those of Mach, Kirchhoff, and Hertz. Finally, the modern trend toward eliminating the concept of force from the conceptual scheme of physical science is fully analyzed.

Needless to say, within the limits of a moderate-sized book it would be impossible to outline the historical development of the individual forces discussed in physics: gravitational, electrostatic, magnetic, frictional, viscous, molecular, nuclear, and so on. A whole library of the size of the *Encyclopædia Britannica* would be insufficient for that purpose. Since classical physics, at the

[1] "Manifestum est enim, quod velocitas motus est ex victoria potentiae moventis super mobile." St. Thomas Aquinas, *In octo libros De Physico Auditu sive Physicorum Aristotelis commentaria* (M. D'Auria Pontificius Editor, Naples, 1953), liber VII (lectio nona), p. 423.

beginning of the present century, recognized essentially only the two forces of gravitation and electromagnetism ("Lorentz forces") and since, in addition, the laws of electromagnetic forces, in general, were formed as far as possible on the analogy of gravitational forces (in spite of their intrinsic polarity characteristics), gravitation more than any other type of forces has been chosen as the most important example of force in the present volume. This selection, moreover, is also justified by the intimate historical connection of the concept of force in classical physics with that of gravitational attraction, although of course the ultimate origin of the concept lies in the muscular sensation associated with push and pull.

As the present research is both a historical treatment and a critical analysis, the author adopted the method of documentary evidence by quoting extensively from the relevant source material. Not only can the reader thus conveniently check the contentions of the author, as far as the latter's interpretation of the sources is concerned, but the reader can also easily differentiate between the historical material as such and the critical comments and personal opinions of the author. Although quotations are sometimes considered as crutches for the intellectually imbecile, it is believed that in the present case citations of original source material, often not easily accessible, will enhance the value of the study.

The book is addressed not only to the professional historian of science but also to the physicist engaged in the historical and epistemological foundations of the conceptual apparatus of his discipline, and to the philosophically minded reader who is interested in the basic problems of science.

Numerous libraries, in different continents, have been consulted. The E. DeGolyer Collection in the History of Science and Technology at the University of Oklahoma, under the efficient curatorship of Professor Duane H. D. Roller, became, however, the author's headquarters for the final rendering of his research. The author wishes to apologize for the fact that, owing to the diversity of libraries consulted, sometimes the same work is quoted in different editions.

The source material is translated into English wherever the knowledge of the sources is thought essential for the understanding of the context. As far as possible, available English translations have been referred to. Only in a few cases, mainly for the use of the scholar, has the original text in Latin, French, or German been quoted.

It is a pleasure to express my gratitude to Professor Roller, who has been most generous with his talent and time in criticizing the manuscript; to my other friends and colleagues at the Department of Physics of the University of Oklahoma, Professor R. G. Fowler, chairman, and to Mr. Harry Starr of the Lucius N. Littauer Foundation, New York, for their advice and encouragement; to Mr. Joseph D. Elder, Science Editor, Harvard University Press, for his fine editing of the manuscript.

I wish to dedicate the present volume to the memory of E. DeGolyer

Department of Physics MAX JAMMER
Hebrew University
Jerusalem, Israel

CONTENTS

CHAPTER 1

THE FORMATION OF SCIENTIFIC CONCEPTS

The purpose of the present study is a presentation of the historical development of the concept of force in physical science. Though recognized as one of the fundamental and primordial notions in physical theory, the concept of force, it seems, has never heretofore been the object of a comprehensive historical analysis and critical investigation. In general, the concept is taken for granted and considered as sanctioned by its successful applications. In fact, standard textbooks, and even elaborate treatises, present little information, if any, on the nature of this concept; its problematic character is completely ignored in the maze of practical utilizations.

The argument, often heard, that the scientist is little concerned with the history of the concepts which he applies in his work, has lost much of its strength in view of the importance that modern physics attaches to concept-formation. Once held as a monopoly of antiquarian historians of science and of pedantic epistemologists, the problem of concept-formation in scientific theory gained vital importance and notable prominence in modern research.

The study of the historical aspects of concept-formation in physical science is admittedly no easy task. In addition to a thorough historical and philological training, necessary for a skillful command of the source material, it requires comprehension of physical theory, instrumental for the critical comparison and interpretation of the sources under discussion and indispensable for the evaluation of their significance for science as a whole.

A serious difficulty in the study of the development of a scientific concept lies in the necessarily inherent vagueness of its definition. This complication arises from the fact that the concept in question finds its strict specification only through its exact definition in science. This definition, however, historically viewed, is a rather late and advanced stage in its development. To limit the discussion to the concept thus defined means to ignore a major part of its life history. The history of a concept has not yet run its course, it is true, even once it has achieved such a "defined" position, since it attains its complete meaning only through the ever-increasing and changing context of the conceptual structure in which it is placed. However, from the standpoint of the history of ideas, the most interesting and important part of its biography is passed, namely the period of its vigor and creative contribution to the advancement of scientific thought. When studying the development of a scientific concept one has, therefore, to cope with an essential vagueness of definition of the subject under discussion and one faces equally the danger of either drawing the limits too narrow or too wide.

As a result of modern research in physics, the ambition and hope, still cherished by most authorities of the last century, that physical science could offer a photographic picture and true image of reality had to be abandoned. Science, as understood today, has a more restricted objective: its two major assignments are the description of certain phenomena in the world of experience and the establishment of general principles for their prediction and what might be called their "explanation." "Explanation" here means essentially their subsumption under these principles. For the efficient achievement of these two objectives science employs a conceptual apparatus, that is, a system of concepts and theories that represent or symbolize the data of sense experience, as pressures, colors, tones, odors, and their possible interrelations. This conceptual apparatus consists of two parts: (1) a system of concepts, definitions, axioms, and theorems, forming a hypothetico-deductive system, as exemplified in mathematics by Euclidean geometry; (2) a set of relations linking certain concepts of the hypothetico-deductive system with certain data of sensory experience. With the aid of these relations, which may be called "rules of interpretation" or "epistemic correlations,"[1] an association is set up, for instance, between a black patch on a photographic plate (a sensory impression) and a spectral line of a certain wavelength (a conceptual element or construct[2] of the hypothetico-deductive system), or between the click of an amplifier coupled to a Geiger counter and the passage of an electron. The necessity for physical science of possessing both parts as constituents results from its status as a theoretical system of propositions about empirical phenomena. A hypothetico-deductive system without rules of interpretation degenerates into a speculative calculus incapable of being tested or verified; a system of epistemic correlations without a theoretical superstructure of a deductive system remains a sterile record of observational facts, devoid of any predictive or explanatory power.

The adoption of rules of interpretation introduces, to some extent, an arbitrariness in the construction of the system as a whole by allowing for certain predilections in the choice of the concepts to be employed. In other words, arbitrary modifications in the formation of the conceptual counterparts to given sensory impressions can be compensated by appropriate changes in the epistemic correlations without necessarily destroying the correspondence with physical reality. In consequence of this arbitrari-

[1] Rudolf Carnap, who, perhaps in analogy to the Kantian conception of a "transcendental scheme," stressed for the first time the importance of these relations, called them *"phenomenal-physikalische Zuordnungen."* See Rudolf Carnap, "Ueber die Aufgaben der Physik," *Kantstudien,* vol. 28 (1923), p. 90. The term "epistemic correlation" is employed by F. S. C. Northrop in his *Logic of the sciences and humanities* (Macmillan, New York, 1947), p. 119.

[2] Cf. Henry Margenau, *The nature of physical reality* (McGraw-Hill, New York, 1950), p. 69.

ness, scientific concepts "are free creations of the human mind and are not, however it may seem, uniquely determined by the external world."[3]

When science attempts to construct a logically consistent system of thought corresponding to the chaotic diversity of sense experience, the selection of concepts as fundamental is not unambiguously determined by their suitability to form a basis for the derivation of observable facts. In the first place, some element of contingency is introduced by the somehow fortuitous sequence of experimentation and observation, an idea recently emphasized by James Bryant Conant: "It seems clear that the development of our modern scientific ideas might have taken a somewhat different course, if the chronological sequence of certain experimental findings had been different. And to some degree, at least, this chronology can be regarded as purely accidental." [4] In the second place, a certain climate of opinion, conditioned by subconscious motives, is responsible to some extent for the specific character of the basic conceptions or primitive concepts. It is a major task of the historian of science to study this climate of opinion prevailing at a certain period and to expose the extrascientific elements responsible for the finally accepted choice of those concepts that were to play a fundamental role in the construction of the contemporaneous conceptual apparatus. The history of science can often show in retrospect how alternative concepts have, or could have, been employed at the various stages in the development of the physical sciences in a provisionally satisfactory manner.

Let us illustrate this point by an example that has some importance for our subject: the Jaina physics[5] of ancient Indian philosophy. The Jainas, followers of Jina (Vardhamāna), an elder contemporary of Buddha, developed a realistic and relativistic

[3] Albert Einstein and Leopold Infeld, *The evolution of physics* (Simon and Schuster, New York, 1938), p. 33.

[4] J. B. Conant, "The citadel of learning," *Yale Review 45*, 56 (1955).

[5] See, for example, Heinrich Zimmer, *Philosophies of India* (Bollingen Series XXVI; Pantheon, New York, 1951).

atomistic pluralism (*anekāntarāda*), without the slightest allusion to a concept of force, in contrast to Western science in which the idea of force plays, as we shall see later in detail, a fundamental role. In the Jaina physics, the category of *ajīva* is subdivided into matter (*pudgala*), space (*ākāsha*), motion (*dharma*), rest (*adharma*), and time (*kāla*). *Dharma* and *adharma* designate the conditions of movement and of rest respectively. Being formless and passive, they do not generate motion or arrest it, but merely help and favor motion or rest, like water, which is instrumental for the motion of a fish, or like the earth, which supports objects that rest on it. Essentially it is "time" that originates "activity" (*kriya*) and "change" (*parināma*), and it does so without becoming thereby some kind of a dynamic agent, something equivalent to the concept of force in Western thought. A more familiar, but less striking, example of a conceptual scheme that does not employ the notion of force is, of course, the physics of Descartes, which, at least in the view of its propounder, was based only on purely geometrico-kinematic conceptions in addition to the notion of impenetrable extension.

Numerous factors compel the scientist to revise constantly his conceptual construction. Apart from general cultural predispositions, conditioned by specific philosophical, theological, or political considerations, the three most important methodological factors calling for such revisions seem to be: (1) the outcome of further experimentation and observation, introducing new effects hitherto unaccounted for; (2) possible inconsistencies in the logical network of derived concepts and their interrelations; (3) the search for maximum simplicity and elegance of the conceptual construction. In most cases it is a combination of two of these factors, and often even the simultaneous consideration of all of them, that leads to a readjustment or basic change of the conceptual structure. A well-known example is the Michelson-Morley experiment which revealed the independence of the velocity of light with respect to the motion of the earth, a phenomenon unaccounted for and inconsistent with the existing ether theory at the turn of the last century; this effect could have been fitted into the conceptual scheme of that time by certain *ad hoc* assumptions (the Lorentz contraction) which would, however, lead to serious complications and violate the principle of simplicity. Einstein's ingenious reinterpretations of space and time, expounded in his special theory of relativity, are essentially a revision of the conceptual apparatus of classical mechanics.

A modification of the prevailing conceptual apparatus has not always, of course, to be so drastic and revolutionary as Einstein's theory of relativity. A common feature of great importance for the historian of scientific concept-formation is the process of "redefining" a concept and changing thereby its status and position in the logical texture of the system. A classical example of this process of redefinition is the concept of temperature. Originally taken as a qualitative expression of heat sensation, it became a quantitative notion when defined as a state of matter measured by the scale reading on a mercury thermometer. When it became apparent, in the further development of this concept, that "temperature" thus defined depended on certain properties of the thermometric substance, it was redefined by the introduction of the so-called "absolute" scale in thermodynamics. Thus it became finally incorporated into a larger and more comprehensive set of relations, forming an integral part of the kinetic theory of matter. It is obvious that by this process the historically and psychologically posterior concept (in the case of "temperature" the kinetic energy of a gas molecule) is treated as the systematically and logically prior, more basic and more fundamental notion.

Not only can concepts which have once been considered as primitive, that is, as basic for nominal definitions of derived concepts, undergo the process of redefinition and thus become derived concepts; although rarely encountered in the history of scientific concept-formation, a concept originally classified as derived may, at a later stage, be chosen as primitive for the redefinition of other concepts. Velocity is generally regarded in classical mechanics as a derived concept, namely, as the ratio of the distance s to the time t, or in the limit, as ds/dt, while distance and time are considered primitive concepts. It would, however, be quite conceivable to formulate a consistent theory of motion by taking the concepts of time t and velocity v as primitive, velocity being measured directly by some kind of speedometer, and to regard distance as a derived quantity according to the defining formula $s = t \cdot v$, or more generally $s = \int v \cdot dt$. In fact, modern astronomy, at least to some extent, does this consistently. The assignment of physical dimensions to these concepts is, of course, no obstacle, since their choice, as is shown conspicuously in electromagnetic theory, is wholly arbitrary and can be chosen in complete conformance with the selection of primitive concepts.

As to the concept of force, taken originally in analogy to human will power, spiritual influence, or muscular effort, the concept became projected into inanimate objects as a power dwelling in physical things. Omitting at present some intermediate stages, the concept of force became instrumental for the definition of "mass," which in its turn gave rise to the definition of "momentum." Subsequently classical mechanics redefined the concept of force as the time rate of change of momentum, excluding thereby, at least prima facie, all animistic vestiges of earlier definitions. Finally, "force" became a purely relational notion, almost ready to be eliminated from the conceptual construction altogether.

Now, a serious question may be asked: does not the recurrent process of redefinition imbue our concept for the most part with a succession of essentially new meanings, so that we should no longer be justified in regarding the various phases as belonging to the same concept? The operationalist who claims that the meaning of a concept is identical with the procedure for measuring it would certainly object to having these various phases treated as different modifications of the same notion; the realist, on the other hand, as well as the advocate of the theory of convergency,[6] for whom scientific statements are more than a complex of mere conventions, would most probably raise no objection. For the historian of science, however, this problem is only a question of words. Not being primarily interested in the problem of reality, that is, in the question how far the internal structure of the hypothetico-deductive system of science reflects or transcribes a possible real substratum underlying the undifferentiated continuum of sensory impressions, the student of the history of ideas stands here on the same footing as the active scientist in his laboratory. Whether his research is a study of the development of one single concept or of a chain of related notions is a completely irrelevant question for him. In other words, whether the various definitions aim at one and the same *definiendum* as part of the reality which transcends consciousness, or whether each modification of the concept has to be regarded as an independent element of the logical system, is a problem to be left for the metaphysician.

Even in the case of a so-called "formalistic"[7] and contextual method of definition, a concept-formation process that is still more problematic in this respect, the attitude of the scientist is the same. In this process the formation of a concept arises from the constancy of certain experimental relations, whereby the constant value obtained is given a special name. Mach's well-known definition of "mass" is an important example: when two bodies, denoted by the subscripts 1 and 2, are acting on each other under the same external conditions, the constant (negative) inverse ratio of their mutually induced accelerations, $(-a_2/a_1)$ is given a new name, the "relative mass" of the two bodies or, more precisely, the mass of the first body relative to the second. If, in

[6] See, for example, Bernhard Bavink, *The anatomy of modern science* (Bell, London, 1932), p. 235.

[7] See, for example, Bernhard Bavink, "Formalistisches und realistisches Definitionsverfahren in der Physik," *Z. physik. chem. Unterricht 31*, 161 (1918).

particular, the second body is a certain standard body ("standard mass"), then the "relative mass" becomes the "mass" of the first body (with subscript 1). It is this highly sophisticated notion of mass that plays the important role in modern physics, both in quantum mechanics (in the determination of the masses of elementary particles) and in the theory of relativity (the dependence of mass on velocity). The earlier, unsophisticated, primitive notion of "*moles*," "quantity of matter," the mass concept of Kepler and Galileo and to some extent even that of Newton, although exhibiting a much closer linkage to elements of sense impressions, would hardly be compatible with the context in which the more refined, essentially relational, Machian concept is integrated. It is important to note that a formalistic definition need not necessarily be a redefinition; the concept of entropy is an example.

The problem of identifying the central notion in the course of the various phases of a development can be a most complicated query. It seems easy to follow the changing modifications of the concept of the electron since Stoney[8] introduced the term in 1874, and to show how this term underwent various transformations until it attained its present status as the name of a fundamental particle in quantum mechanics; it is, however, much harder to decide whether the medieval concept of impetus is the forerunner of the notion of momentum of classical mechanics; it is still more complicated to determine whether the Aristotelian idea of *dynamis* is the antecedent of the Newtonian concept of force.

The identification of the central notion, in our case, is further complicated by the great obscurity and confusion involved in the terminology. The term "force" or its equivalent in other languages has been used in a great variety of different significations. Needless to say, we are not concerned with the figurative sense implied in expressions like "force of example," "force of habit,"

[8] G. J. Stoney, "On the physical units of nature," *Phil. Mag.* [5] *11*, 384 (1881).

"police force," "economic forces," and so on. On the other hand, an expression like "forces of nature," if used in a scientific sense, may have some relevance for our investigation.

But even when used as a scientific technical term, "force" may imply different significations according to the context in which it was employed. For instance, a critical analysis would show that in the following sentence, written by Herbert Spencer, each usage of the term "force" implies a different meaning: "By conflict with matter, a uniform force is in part changed into forces differing in their directions; and in part it is changed into forces differing in their kinds." [9] In Descartes' mechanics, which, as we shall see, is essentially only a theory of percussion, the third law of motion states that a body in motion in a straight line, when colliding with another body endowed with less "force" (*vis*), continues its motion in the original direction, losing as much motion as the other body absorbs. But if the "force" of the second body is greater than the "force" of the first, the latter will change its direction without losing its motion.[10] It is obvious that "force" (*vis*) in this context denotes what we call "quantity of motion" or "momentum," that is, mass times velocity. Descartes, of course, is not to be blamed for this discrepancy. His investigation was leading him into realms of thought which, theretofore unexplored, suggested ideas for the expression of which no words were yet available. What was, therefore, more natural than to take a word from ordinary Latin, a word that in particular as yet had no technical connotation, and to use it as a *terminus technicus*, especially as its ordinary meaning seems to come rather close to the operational interpretation of the concept under discussion? Descartes' use of the term "force" (*vis*) for "mass times velocity"

[9] Herbert Spencer, *First principles* (Appleton, New York, 1895), p. 432.

[10] "Tertia lex naturae haec est: ubi corpus quod movetur alteri occurrit, si minorem habeat vim ad pergendum secundum lineam rectam, quam hoc alterum ad ei resistendum, tunc deflectitur in aliam partem . . ." Renati Descartes, *Principia philosophiae*, pars secunda, cap. XL, in *Oeuvres de Descartes*, ed. Charles Adam and Paul Tannery (Cerf, Paris, 1905), vol. 8, p. 65.

is thus wholly legitimate, although subsequent scientific terminology took a different path.

The situation, however, becomes complicated when the same author employs the same term in different meanings. In Descartes' letter to Mersenne of November 15, 1638, the term "force" is employed as follows: "Vous avez enfin entendu le mot de force au sens que je le prens, quand je dis qu'il faut autant de force, pour lever un poids de cent livres à la hauteur d'un pied, qu'un de cinquante à la hauteur de deux pieds, c'est à dire qu'il y faut autant d'action ou autant d'effort." [11]

The passage itself makes it quite clear that "force," in this context, is our concept of "work." This is also the meaning of the term "force" in the beginning of Descartes' short treatise, "Explication des engins par l'ayde desquels on peut avec une petite force lever un fardeau fort pesant." Here he says: "L'invention de tous ces engins n'est fondée que sur un seul principe, qui est que la mesme force qui peut lever un poids, par example, de cent livres à la hauteur de deux pieds, en peut aussy lever un de 200 livres, à la hauteur d'un pied, ou un de 400 à la hauteur d'un demi pied, et ainsy des autres, si tant est qu'elle luy soit appliquée." [12]

As early as 1743 d'Alembert protested against this confused and indiscriminate use of the term "force" when he said: "When we speak of the 'force of a body in motion' either we form no clear idea of what this expression means or we understand by it only the property which moving bodies have of overcoming obstacles encountered in their path or of resisting them." [13] The nomenclature became even more ambiguous during the nineteenth century when the term "force" was used regularly also to denote

[11] *Oeuvres de Descartes,* ed. Adam and Tannery (Cerf, Paris, 1898), vol. 2, p. 432.

[12] *Oeuvres de Descartes,* ed. Adam and Tannery (Cerf, Paris, 1897), vol. 1, p. 435.

[13] J. L. d'Alembert, *Traité de dynamique* (David, Paris, 1743), "Discours préliminaire," p. xvi (preface).

our present-day notion of "energy" and "work," the reason being Leibniz's coinage of *vis viva* (our "kinetic energy").

How impeditive such confused terminology can be, not only for the historian of science, but also for the contemporaries of the author who uses the term, is best shown by the fate of Julius Robert Mayer's paper on conservation of energy, entitled "On quantitative and qualitative determination of forces,"[14] which he submitted on June 16, 1841, for publication to Johann Christian Poggendorff, the editor of *Annalen der Physik und Chemie*. Poggendorff rejected this important piece of research, which Liebig accepted in 1842 for publication in his *Annalen der Chemie und Pharmazie* under the title "Observations on the forces of inanimate nature."[15] The publication, however, attracted no attention at all. As Mach relates,[16] even Jolly when visiting Mayer in Heidelberg finally understood the implications and arguments of his host only after long discussion and "considerable explanation." In 1845 Mayer published his famous article, "Organic motion with reference to metabolism,"[17] in which he says: "Es gibt in Wahrheit nur eine einzige Kraft. In ewigem Wechsel

[14] J. R. Mayer, "Ueber quantitative und qualitative Bestimmung der Kräfte." Cf. E. Dühring, *Robert Mayer, der Galilei des neunzehnten Jahrhunderts* (Schmeitzner, Chemnitz, 1880).

[15] J. R. Mayer, "Bemerkungen über die Kräfte der unbelebten Natur," *Annalen der Chemie und Pharmacie von Wöhler und Liebig 42,* 233 (1842); reprinted in J. R. Mayer, *Die Mechanik der Wärme* (Cotta, Stuttgart, 1867), pp. 1–12.

[16] "It was only after considerable explanation that Jolly found out what Mayer wanted to say." Ernst Mach, "On the part played by accident in invention and discovery," *Monist 6,* 171 (1896); reprinted in: Ernst Mach, *Popular scientific lectures,* trans. by T. J. McCormack (Open Court, LaSalle, Illinois, 1943), p. 274.

[17] J. R. Mayer, "Die organische Bewegung in ihrem Zusammenhang mit dem Stoffwechsel. Ein Beitrag zur Naturkunde," reprinted in Mayer, *Die Mechanik der Wärme,* p. 13–126. The article opens with the statement: "Soll eine ruhende Masse in Bewegung gesetzt werden, so ist dazu ein Aufwand von Kraft erforderlich. Eine Bewegung entsteht nicht von selbst; sie entsteht aus ihrer Ursache, aus der Kraft. Ex nihilo nil fit. Ein Object, das, indem es aufgewendet wird, Bewegung hervorbringt, nennen wir Kraft. Die Kraft, als Bewegungsursache, ist ein unzerstörliches Object."

kreist dieselbe in der toten wie in der lebenden Natur. Dort und hier kein Vorgang ohne Formveränderung der Kraft."

That this homonymic use of the term "force" (*Kraft*) led to serious confusion can also be seen from Hermann von Helmholtz's classical paper, "Ueber die Erhaltung der Kraft,"[18] read at the age of twenty-six before the Physical Society of Berlin on July 23, 1847. Misunderstood, it was looked upon as a most fantastic speculation and had the same fate as Mayer's article, being also rejected for publication by Poggendorff. The controversy between Clausius and Dühring on the one hand, who attacked the paper, and Du Bois-Reymond and Jacobi on the other, who defended its implications, is an excellent illustration of the confusion evoked by the ambiguous use of the term. That these vague expressions came into common use also in nonscientific language is not surprising. Let us quote a passage in which the German humorist Wilhelm Busch, who started his career as an engineer, describes a box on the ear:

Hier strotzt die Backe voller Saft,
Da hängt die Hand gefüllt mit Kraft;
Die Kraft infolge der Erregung
Verwandelt sich in Schwungbewegung;
Bewegung, die mit schnellem Blitze
Zur Backe eilt, wird hier zur Hitze.
Die Hitze aber durch Entzündung
Der Nerven brennt als Schmerzempfindung
Bis auf den tiefsten Seelenkern,
Und dies Gefühl hat keiner gern.
Ohrfeige nennt man diese Handlung
Der Forscher nennt es Kraftverwandlung.[19]

After these perhaps somewhat intimidating remarks on the difficulties to be expected in a careful analysis of concepts from the historical point of view, it seems to be appropriate, on the

[18] Hermann von Helmholtz, "Ueber die Erhaltung der Kraft," *Ostwald's Klassiker,* No. 1 (Engelmann, Leipzig, 1889).

[19] Wilhelm Busch; *Balduin Bählamm, der verhinderte Dichter* (Bassermann, Munich, 1883), chap. 6.

other hand, to stress also the great importance of our subject both for science and for philosophy. In the first place, "force" is one of the first concepts, and, in many textbooks of physics, the very first nonmathematical concept that the student of science encounters. In the course of his study he meets this concept over and over again: he studies gravitational force, electromagnetic force, frictional and viscous forces, cohesive and adhesive forces, elastic and chemical, molecular and nuclear forces. If not clarified and subjected to critical analysis these concepts are liable to be misunderstood as mystic entities or even occult qualities playing a central role in present-day physics. Charles Sanders Peirce, who realized this situation, when interpreting the concept from his pragmatic point of view, said that force is "the great conception which, developed in the early part of the seventeenth century from the rude idea of a cause, and constantly improved upon since, has shown us how to explain all the changes of motion which bodies experience, and how to think about physical phenomena; which has given birth to modern science, and changed the face of the globe; and which, aside from its more special uses, has played a principal part in directing the course of modern thought, and in furthering modern social development. It is, therefore, worth some pains to comprehend it."[20]

Indeed, the clear comprehension of mechanical force and the conscious incorporation of it into the basic structure of physics can be regarded as the beginning of modern science. Whereas Aristotelian-Ptolemaic science was predominantly a system of essentially geometrico-kinematic conceptions, the "new science" was based on Newtonian dynamical interpretations. A critical evaluation of the importance of the concept of force for modern science is therefore undoubtedly of the greatest significance for an understanding of the development of modern science.

But more than this is involved. "Force" has a unique position among all possible basic concepts in physical science since it may

[20] C. S. Peirce, *Collected papers,* ed. Charles Hartshorne and Paul Weiss (Harvard University Press, Cambridge, 1934), vol. 5, p. 262.

be regarded as having a direct relation to the concept of cause. Indeed, many students of the problem, and foremost among these the Kantian school of thought, consider "force" the exact physical formulation of "cause" and causality. According to this point of view, natural science relates all phenomena in nature to certain substrata, the phenomena being conceived as their effects. Now, in the consistent performance of this operation the scientific conditions of this substratum have to be formulated in such a way that causal connection is maintained and preserved throughout. The concept of substance thus derives from the empirical applications of the principle of causality. The notion of substance thus formed is used, in its turn, to deduce from it the particular causal connections. Causality, thus attached to substance, is called "force" and the substance to which the actions of this force are referred is regarded as the "carrier" of the force.

Owing to this particular position, the concept of force became the main object of severe attacks from the quarters of positivism; for positivistic thought contended that the elimination of the concept of force from physics would lead to the emancipation of science as a whole from the bondage of causality, one of the "most obstinate remnants of prescientific fetichism."

Last, but not least, a critical exposition of the development of the concept of force forms an important chapter in the history of ideas, since it reflects the constant change in intellectual attitude throughout the ages.

CHAPTER 2

THE CONCEPTION OF FORCE IN ANCIENT THOUGHT

Science, as a whole, is unquestionably a gradual and continuous outgrowth from ordinary everyday experience. It would therefore be only natural to assume that scientific concepts, too, have their ultimate origin and foundation in the conceptions of everyday experience. This statement does not contradict the contention that scientific concepts are free creations of the human intellect, since, in the first place, the concepts of ordinary prescientific thought are themselves the outcome of arbitrary decompositions of the coherent and continuous substratum of sensory experience. Second, science as a technical activity never tries purposely to detach itself from conceptions formed in ordinary experience; on the contrary, scientific conceptions, though frequently the result of spontaneous intuition, tend to be molded, as far as possible, in analogy to the conceptions of ordinary experience. It is no exaggeration to contend that the role of analogy is very important for the progress of knowledge, reducing the unknown and the strange to the terms of the familiar and the known. In this sense all cognition is recognition. When studying the historical formation of a concept, one should remember that the metaphor is a powerful agent in the evolution of language as well as of science; it is instrumental in the transference of a word from its ordinary meaning to the designation of a specific concept as a defined construct within the conceptual scheme of science.

It has often been pointed out that scientific terminology, in particular in ancient science, supplies some information about the possible ways of transmission of scientific ideas from one civilization to another; similarly, the ancient nomenclature of fundamental scientific concepts throws some light on the history of their origin and evolution. Since even the most technical jargon is based ultimately on conversational language, philological research proves an important factor in the historical study of scientific notions.

The idea of force, in the prescientific stage, was formed most probably by the consciousness of our effort, spent in voluntary actions, as in the immediate experience of moving our limbs, or by the consciousness of the feeling of a resistance to be overcome in lifting a heavy object from the ground and carrying it from one place to another. Clearly, "force," "strength," "effort," "power," and "work" were synonymous, as they still are today in ordinary unsophisticated language. The injection of our personal experience into the external environment, characteristic of the animistic stage in the intellectual growth of mankind, led to a vast generalization of the concept of force: trees, rivers, clouds, and stones were endowed with force and were regarded as centers of power. For what is active was thought to be alive, and an object, animal or material, being alive, was conceived as having within it the same sort of force that man recognized in himself. Moreover, things of nature that seemed to be powerful were not only endowed with an anthropopathic nature; they also became objects of fear and reverence.

It is tempting to compare these generally accepted ideas on the origin of the concept of force with the psychological formation of this concept in the mind of the child. According to leading schools in modern psychology, the order of this concept-formation in the mind of the child is the exact reverse of the order outlined above. Whereas the idea of force, as explained above, is carried over from the inner experience and consciousness of the subject, the child seems to attribute forces to the objects of his external environment before he finds in himself the "ego" as the cause of his own force. "In the early stages during which the child's realism, i.e., his ignorance of his own ego reaches its highest point, childish dynamism is complete: the universe is peopled with living and substantial forces to a degree that adult common sense would find it hard to imagine. During the later stages, as the child gradually becomes conscious of the inner world and of the specificity of his ego, dynamism is ousted from the child's conception of the world by a more mechanical way of thinking."[1]

Whatever the order of transference, both for the prescientific and for the infantile mind the dynamism of the external nature is a panpsychism in which every object is endowed with a force *sui generis,* unacquired and untransmissible. The later belief in occult powers, inherent in inanimate objects, the belief in amulets, the "mana" of Oceania, and similar notions, are retraceable to this pandemonic interpretation of nature. The familiar expression of "forces of nature" is still reminiscent of this outlook on nature.

With the progressive organization of early society into urban civilization the concept of a capricious interplay of forces behind the ever-shifting phantasmagoria changed into the idea of a systematized hierarchy of forces in nature; eventually "force" as such was personified into a spirit or a god of overwhelming power. Such personification was characteristic of ancient mythology which, as the only body of systematized thought of those times, was not only the cosmology but also the "physics" of the prescientific stage. Now, what was the mythological concept of force?

The abstract concept of force, as a notion of divinity, can be traced back in ancient Egypt to the nineteenth dynasty[2] and plays an important role in later texts of the demotic literature. The term "nḫt," denoting the divine personification of "force" in

[1] Jean Piaget, *The child's conception of physical causality* (Humanities Press, New York, 1951), p. 128.

[2] Papyrus Harris, 500.

ancient Egypt, has been studied by Hess[3] and later by F. L. Griffith.[4] Wilhelm Spiegelberg in his paper on "The Egyptian divinity of 'force'" says explicitly: "This abstract concept of 'force' with its attribute 'divine' as stated explicitly in the inscription was conceived as a personal deity."[5] As an illustration he refers to the hieroglyphic rendering of the demotic name "Pn-n-nḫt-w" (Fig. 1), in which the figure with the knife repre-

Fig. 1. The name "Pn-n-nḫt-w" in hieroglyphics.

sents the concept of divine force ("Gotteskraft"). H. Bonnet concurs with Spiegelberg in the interpretation and rejects the possibility of identifying "nḫt" with a daemonic or angelic intermediary. "Instead it has to be interpreted as forces, conceived as personal beings, emanating from the deity."[6]

The nature of this concept of "divine force" in later demotic literature is well illustrated in the Setne story, in which Toth entreats Re to return the book of magic which has been stolen by Ne-nefer-ke-Ptah; whereupon the sun-god fulfills this request by "sending down a 'divine force' from heaven," charged with the prosecution of the robber. This example shows that "force" as personified by "nḫt" is not only force in the sense of violence and ferocity, but includes already at this early stage an element of order and morality.

[3] J. J. Hess, *Der demotische Roman von Stne Ha-m-us* (Leipzig, 1888), p. 73.

[4] F. L. Griffith, *Stories of the high priests of Memphis* (Clarendon Press, Oxford, 1900), pp. 26, 109.

[5] "Dieser abstracte Begriff 'Kraft' mit dem in der Schrift ausgedrückten Zusatz 'göttlich' wird durchaus als persönliche Gottheit empfunden." Wilhelm Spiegelberg, "Die ägyptische Gottheit der 'Gotteskraft,'" *Zeitschrift für ägyptische Sprache und Altertumskunde 57,* 148 (1922).

[6] "Vielmehr wird man sie als aus der Gottheit emanierende Kräfte anzusprechen haben, die persönlich gedacht sind." Hans Bonnet, *Reallexikon der ägyptischen Religionsgeschichte* (De Gruyter, Berlin, 1952), p. 254.

A similar development can be seen in Mesopotamian civilization. The population of ancient Mesopotamia, much more than that of ancient Egypt, was subjected to fitful and unpredictable "forces of nature." Thunderstorms whose "dreadful flares of light cover the land like cloth," and devastating winds and floods "which shake the heavens and cause earth to tremble," dealt destruction to land and man. Force, for the inhabitant of the Valley between the Two Rivers, is personified by Enlil, the god of storm.[7] Enlil stands for inconstancy, vicissitude, and dynamic action, in contrast to the sky-god Anu, one of the most ancient divinities, whose name is found on the earliest inscriptions, Anu meaning "the expanse of heaven" and symbolizing through space and static extension the principle of immutability and permanence.

> Thus, in the society which the Mesopotamian universe constitutes, Anu represents authority, Enlil force. The subjective experience of the sky, of Anu, is . . . one of majesty, of absolute authority which commands allegiance by its very presence . . . Not so with Enlil, the storm. Here, too, is power; but it is the power of force, of compulsion. Opposing wills are crushed and beaten into submission. In the assembly of the gods, the ruling body of the universe, Anu presides and directs the proceedings. His will and authority, freely and voluntarily accepted, guide the assembly as a constitution guides the actions of a law-making body. Indeed, his will is the unwritten living constitution of the Mesopotamian world state. But whenever force enters the picture, when the cosmic state is enforcing its will against opposition, then Enlil takes the centre of the stage.[8]

Again, Enlil represents force not only in the sense of brutality and might; he also stands for force as the ordering element in the universe, as the regulative norm against the chaos.

Force and power are frequent attributes of the God of the Bible. In Exodus 32:11 we read: "And Moses besought the Lord his God, and said, Lord, why doth thy wrath wax hot against

[7] The cult of Enlil is probably of Sumerian origin. Later he is frequently referred to as Bel, to be displaced, subsequently, by Marduk.

[8] Thorkild Jacobsen, "Mesopotamia," in H. Frankfort and others, *Before Philosophy — The Intellectual Adventure of Ancient Man* (Penguin Books, Baltimore, 1949), p. 156.

thy people, which thou hast brought out of the land of Egypt with great power, and with a mighty hand?" But God, for the Children of Israel, is no longer a neutral power, a force of nature, blind and capricious; God is the force of morality and the will for the good.

O Lord God, thou hast begun to shew thy servant thy greatness, and thy mighty hand; for what God is there in heaven or in earth, that can do according to thy works, and according to thy might? [9]

The voice of the Lord is powerful;
The voice of the Lord is full of majesty.
The voice of the Lord breaketh the cedars;
Yea, the Lord breaketh the cedars of Lebanon.
The Lord will give strength unto his people;
The Lord will bless his people with peace.[10]

God is the source of force and power, but force of good will and of justice.

"Clouds and darkness are round about him; righteousness and judgment are the habitation of his throne." [11] Consistently with the spirit of its prophets, Judaism transformed the idea of force into the notion of an ethical power.

The relation of the Biblical God to the notion of force is clearly manifested by the frequent association of his name with the idea of force in the form of *hail* ("might," 164 passages), *g'vurah* ("power," 13 passages), *coah* ("strength," 8 passages), *os* ("vigor," 21 passages),[12] terms which in the Septuagint are translated mostly by the expression "dynamis." Some etymologists have believed that one of the holy names, "shaddai," derives from the Semitic root "shadda," meaning "to have great power, great force," and translated in the Septuagint as "pantokrator" ("omnipotens").[13] A study of divine epithets in other ancient lan-

[9] Deut. 3:24. [10] Psalm 29:4, 5, 11. [11] Psalm 97:2.

[12] For further reference, see Gerhard Kittel, *Theologisches Wörterbuch zum neuen Testament* (Kohlhammer, Stuttgart, 1950), p. 286.

[13] *Encyclopedia Biblica* (Jerusalem, Bialik Institute, 1950). See also Ibn Esra, *Sepher Hashem* f.4b, commentary to Exod. 6:3. Cf. David Kaufmann, *Geschichte der Attributenlehre in der jüdischen Philosophie des Mittelalters* (Perthes, Gotha, 1877), p. 181.

guages will show the same occurrence. Thus a proper name of an early Arabic deity was "Al-Uzza" ("the mightiest"), which later, most probably, became the common compellation "Al-Aziz" of Allah. Donar-Thor, the acknowledged god of all Teutons, the god who manifests himself in the thunder, represents again force and power when he hurls his axe or his hammer from the sky. As is related by Herodotus,[14] the ancient Persian or Iranian conception of god is that of great forces of nature. Even post-Biblical literature recognizes forces as an outstanding quality of God. On hearing the words, "O God, the Great, the Strong, the Awful, the Mighty, the Powerful, the Bold," Rabbi Hanina silenced the speaker, saying that it would be derogatory to call God by these names for it would be "praising a millionaire for possessing only a hundred thousand." [15]

While one of the supreme tenets in the philosophy of the Bible is the reconciliation of force with righteousness and morality, Greek prescientific thought strives for a compromise between force and destiny. Zeus, the unrivaled ruler of the universe, who wields the awful thunderbolt, represents for the Greek mind the idea of might and power. All the Greek gods were originally departmental powers in charge of the elements, who, as already related by Herodotus,[16] acquired their names and personal attributes by degrees; thus Zeus stood for the all-pervading force in nature. His force is the greatest, greater than the power of all the other gods together. He tells his family in the *Iliad:*

> Then he will see how far I am strongest of all the immortals.
> Come, you gods, make this endeavour, that you all may learn this.
> Let down out of the sky a cord of gold; lay hold of it
> all you who are gods and all who are goddesses, yet not
> even so can you drag down Zeus from the sky to the ground, not
> Zeus the high lord of counsel, though you try until you grow weary.
> Yet whenever I might strongly be minded to pull you,
> I could drag you up, earth and all and sea and all with you,

[14] Herodotus, book I, chap. 131, the text of Canon Rawlinson's translation (Murray, London, 1897), vol. 1, p. 78.

[15] *Megila,* 25a.

[16] Herodotus, book II, chap. 52.

then fetch the golden rope about the horn of Olympos
and make it fast, so that all once more should dangle in mid air.
So much stronger am I than the gods, and stronger than mortals.

Such is the way, it will be pleasing to Zeus, who is too strong,
who before now has broken the crests of many cities
and will break them again, since his power is beyond all others.[17]

Zeus, in the Homeric conception, although the symbol of force and power, is not omnipotent. He, too, is subjected to the supremacy of Moira, the impersonal and unalterable law. Destiny is stronger than Zeus. In later development of Greek thought, however, a union between Zeus and the Fates is aimed at. As much as the union between mind and necessity, in the Orphic tradition, made the creation of the world possible, so the reconciliation of force with destiny assures regularity in the course of events. In the opening scene of Aeschylus' *Prometheus bound,* Zeus is represented by his two ministers, Kratos and Bia, the allegories of Force and Violence. But at the end of the *Eumenides* we read:

There shall be peace forever between these people of Pallas and their guests. Zeus the all-seeing met with Destiny, to confirm it.[18]

Thus we see that the concept of force, since its early inception in the systematic pattern of thought in all ancient civilizations, is closely related to religious ideas, a relation that it maintains throughout the early stages of its development and particularly in the Platonic doctrine of force as an emanation of the world soul, as will be explained in the following chapter. It also plays an important role in the metaphysical foundations of Newtonian dynamics, as we shall see later on.

[17] *The Iliad of Homer,* book VIII, 17, book II, 116, trans. Richmond Lattimore (University of Chicago Press, Chicago, 1951), pp. 182, 79.

[18] Aeschylos, *Oresteia,* trans. Richmond Lattimore (University of Chicago Press, Chicago, 1954), p. 171.

CHAPTER 3

THE DEVELOPMENT OF THE CONCEPT OF FORCE IN GREEK SCIENCE

The early cosmologists, such as Thales, Anaximander, or Anaximenes, conceived nature, the primary substance, as a living being, self-moving and giving birth to individual things.[1] Regarding the substance of the world as organic and immortal, they felt no difficulty about the cause of motion and did not raise the problem of its possible origin. Although Aristotle reports that they believed in eternal motion, it seems that the problem of motion and its origin was not discussed by these cosmologists at all. Besides, it would have been rather rest, the absence of motion, that had to be accounted for by their hylozoistic natural philosophy. Only at a later stage, when the primary element was reduced to the level of corporeal, inanimate matter, could the problem of an external agent, the cause of motion, be raised. In fact, the conception of dynamic causation had its origin in the reaction to Eleatic thought. For Parmenides, what is, is a finite, immovable, indivisible, and continuous plenum; becoming and passing away, motion and change, are nonexistent.

In order to avoid this dilemma, the belief in the essential oneness of existence had to be given up, while change in its constituents and motion of its parts had to be accounted for. Motion,

[1] "Aristotle and Hippias say that Thales attributed souls also to lifeless things, forming his conjecture from the nature of the magnet, and of amber." Diogenes Laertius, *De clarorum philosophorum vitis, dogmatibus et apophthegmatibus libri decem, Graece et Latine,* ed. C. G. Cobet (Firmin-Didot, Paris, 1878), lib. I, cap. 24.

hitherto taken for granted as characteristic of the bodily and inherent in nature, became the object of philosophical analysis. While Parmenides denied the possibility of motion, his opponents had to explain its origin.

One of the earliest dynamic conceptions of nature is Heraclitus' doctrine of opposing tensions, according to which all things, although in appearance most stable, are but the battlefields of antagonistic forces; their stability is only relative, or even illusory; the concept of force is still confined to this inherent antagonism and balanced conflict of opposites in the individual object, each item of existence being the battleground of opposing forces and tensions. "The concord of the universe, like that of a lyre or bow, according to Heracleitus, is resilient if disturbed." [2] Force is not yet considered by him to be a regulative element in the universe. Thus, Helios, the sun, does not transgress his appointed limit, but runs his path immutably and unchangeably. "For the sun never transgresses its limited measures, as Heraclitus says; if it did do so, the Furies, which are the attendants of Justice, would find it out and punish it." [3]

Force as a regulative agent in nature appears, perhaps for the first time in Greek thought, in Empedocles' doctrine of love and strife,[4] and in Anaxagoras' theory of the mind (*nous*). Both doctrines aimed at an explanation of the causes of motion. "What they were feeling after was obviously the later physical conception of force, but it is equally clear that they were still unable to disentangle this completely from that of body." [5] These agents as causes of motion may rightfully be interpreted as "forces," although they were not held as immaterial, but as extended in space and corporeal.

[2] Plutarch, *Moralia,* "De Iside et Osiride," sec. 45, trans. F. C. Babbitt (Loeb Classical Library; Harvard University Press, Cambridge, 1936), vol. 5, p. 109.

[3] *Plutarch's Morals* ("of banishment, or flying one's country"), trans. W. Goodwin (Boston, 1878), vol. III, p. 26.

[4] How far ancient Orphic tradition may be regarded as the source of Empedocles' dynamistic doctrine is difficult to determine.

[5] John Burnet, *Greek philosophy* (Macmillan, London, 1950), p. 70.

By transferring the Parmenidean conception of "being" to the primary elements, these latter are conceived as unchangeable substances. In order to explain "becoming" and "change," Anaxagoras is led to the assumption of a separate moving force, external to matter, some kind of world-forming spirit which operates as a force in the universe. Since man's experience affords only one analogy for incorporeality and for design, and this is the human spirit, Anaxagoras employs the term "mind" (*nous*)[6] to denote this dynamic ordering factor in nature. Primitive matter, prior to the intervention of "mind," is not yet organized and is unmoved. "Mind" is the external agent, the source of impulse of motion and the cause of change and variation. It is, however, not yet spirit as contrasted to matter. It is "the subtlest of all things," represented in a sensible form of a refined kind of matter. Although Anaxagoras' doctrine comes near to recognizing the immateriality of mind and may therefore be regarded as the beginning of the agelong breach between mind and matter, his conception of force as "mind" shows still the characteristics of a corporeal substance. Force is still some kind of fluid substance, though different from all other material things.

Similar considerations apply to Empedocles' conception of force. In order to make "birth" and "decay," "combination" and "separation," comprehensible, Empedocles thinks it necessary to introduce two separate antagonistic forces: "love" (*philotes, philia*) and "strife" (*neikos*). "Love" and "strife," however, were classed with the four elements, earth, water, air, and fire. These elements, according to Empedocles, were mixed among themselves while love, so to say, served as the binding power linking the various parts of existence harmoniously together. When strife, however, invaded the sphere of being from his outside position, love was driven to the center and the four elements became

[6] "Spirit" or "psyche" is here understood in the ancient meaning of "breath-soul." For the etymology of this term, as well as for that of nous, see R. B. Onians, *The origin of European thought* (Cambridge University Press, Cambridge, 1954), pp. 82–83.

alienated and separated from one another, until the reverse process of the expansion of love was reënacted. The periodic influx of strife, accompanied by the contraction of love, and the reverse process have rightly been interpreted as an adaptation of the ancient idea of the world as a breathing organism. As is natural for the founder of a medical school, Empedocles derives the physics of the universe from the physiology of the body. If Empedocles' principle of love can be regarded as the early formulation of physical attraction, the function of love being the production of union, then the concept of attractive forces has its origin in physiology. Indeed, Empedocles, being a naturalist, did not feel the need for a mechanical explanation of the cosmic systole and diastole. As for all early Greek science, the animal organism is, in principle, simpler than any artificial man-made mechanism — Aristotle still explains mechanism by the analogy to animal organism — so for Empedocles the flux and reflux of the blood to and from the heart is logically prior to and the basis of his physiological analogy.

Plato interpreted Empedocles' two agents as attraction and repulsion, stating that their operation is conceived in an alternate sequence, whereas, according to Plato, the same forces operate simultaneously in Heraclitus' conception of nature. Aristotle describes Empedocles' love and strife correctly as having a double character: "The love of Empedocles is both an efficient cause, for it brings things together, and a material cause, for it is a part of the mixture."[7] Similarly says Theophrastus: "Empedocles sometimes attributed efficient powers to love and strife, and sometimes put them on the same footing as the four elements."[8] In fact, apart from being a cause of motion, love, in the sense of Empedocles, is to be taken on the same level as the four elements and coextensive with them. "Fire and water and earth and the boundless height of air, and also execrable hate apart from these,

[7] Aristotle, *Metaphysics* A, 10. 1073 b 3.
[8] Theophrastus, *Phys. Op.* fr. 3 (Dox. p. 477. R.P. 166b).

of equal weight in all directions, and love in their midst, their equal in length and breadth."[9] As everlasting as the four elements, love and strife have an eternal life.

> For even as Love and Hate were strong of yore
> They shall have their hereafter; nor I think
> Shall endless Age be emptied of these twain.
>
> Now grows
> The One from many into being, now
> Even from the One disparting come the Many,—
> Fire, Water, Earth and awful heights of Air;
> And shut from them apart, the deadly Strife
> In equipoise, and Love within their midst
> In all her being in length and breadth the same.
> Behold her now with mind, and sit not there
> With eyes astonished, for 'tis she inborn
> Abides established in the limbs of men.
> Through her they cherish thoughts of love, through her
> Perfect the works of concord, calling her
> By name Delight or Aphrodite clear.
> She speeds revolving in the elements,
> But this no mortal man hath ever learned—
> Hear thou the undelusive course of proof:
> Behold those elements own equal strength
> And equal origin; each rules its task;
> And unto each its primal mode; and each
> Prevailing conquers with revolving time.
> And more than these there is no birth nor end;
> For were they wasted ever and evermore,
> They were no longer, and the great All were then
> How to be plenished and from what far coast?
> And how, besides, might they to ruin come,
> Since nothing lives that empty is of them?—
> No, these are all, and, as they course along
> Through one another, now this, now that is born—
> And so forever down Eternity.[10]

Plato's conception of force is closely related to his metaphysical doctrine of being. As for the early cosmologists motion was an

[9] Hermann Diels, *Die Fragmente der Vorsokratiker* (Weidmann, Berlin, 1922), fragment B 17, vol. 1, p. 230.

[10] *The Fragments of Empedocles,* trans. W. E. Leonard (Open Court, Chicago, 1908), pp. 20, 22.

inherent property of matter, matter, in their view, being a living organism, so for Plato physical reality is endowed with motion because nature has an immortal living soul. Plato, however, in consequence of his metaphysical teachings, has to face the problem of the origin of motion. Whether locomotion or rotation, separation or combination, or any other of the ten different kinds of movement listed in his dialogue "The Laws,"[11] all motion is reduced ultimately to spontaneous motion, the principle of life and soul. Plato's association of motion with soul is reminiscent of Alcmaeon's statement that the soul is immortal because it is forever in motion like the sun, the moon, and the stars. That self-motion is the cause of all possible forms of motion is discussed in detail in "The Laws" (book X):

> *Athenian Stranger:* Then we must say that self-motion being the origin of all motions, and the first which arises among things at rest as well as among things in motion, is the eldest and mightiest principle of change, and that which is changed by another and yet moves other is second.
>
> *Cleinias, a Cretan:* Quite true.
>
> *Ath.:* At this stage of the argument let us put a question.
>
> *Cle.:* What question?
>
> *Ath.:* If we were to see this power existing in any earthly, watery, or fiery substance, simple or compound — how should we describe it?
>
> *Cle.:* You mean to ask whether we should call such a self-moving power life?
>
> *Ath.:* I do.
>
> *Cle.:* Certainly we should.
>
> *Ath.:* And when we see soul in anything, must we not do the same — must we not admit that this is life?
>
> *Cle.:* We must.[12]

For Plato, as for Empedocles, the four elements as such have no source of motion in themselves. However, Plato's conception of being as producing change and itself in turn affected by other causes leads him to contend that nature is endowed with motion, the self-moving principle of life and of soul being the basis of all

[11] *The dialogues of Plato,* trans. Benjamin Jowett (Random House, New York, 1937), vol. 2, p. 634.

[12] *Ibid.,* p. 637.

physical processes. In fact, the manifestation of real things is nothing but power. "My notion would be, that anything which possesses any sort of power to affect another, or to be affected by another, if only for a single moment, however trifling the cause and however slight the effect, has real existence; and I hold that the definition of being is simply power." [13]

Thus, what exists has power; but power means motion which reduces to the principle of self-motion or the soul. "The soul through all her being is immortal, for that which is ever in motion is immortal; but that which moves another and is moved by another, in ceasing to move ceases to live. Only the self-moving, never leaving self, never ceases to move, and is the fountain and beginning of motion in all that moves beside." [14]

Plato's conception of force as something intrinsic in matter because matter has soul conforms with his view that the mutable world is composed of one single universal "this" susceptible of many differentiated "suches"; in principle, it is one substance or one substratum and one universal receptacle which manifests itself through forces and forms as different aspects in the actual "this." "Anything which we see to be continually changing, as, for example, fire, we must not call 'this' or 'that,' but rather say that it is 'of such a nature'; nor let us speak of water as 'this,' but always as 'such.'" [15] The actual and particular differentiation of the one being is realized through the activity of forces, emanating from the world-soul, an idea that through Neo-Platonic interpretations was to have a great influence on the concept of force. The ultimate origin of all forces in nature, according to this interpretation, lies in the hidden world-soul. "The world has received animals, mortal and immortal, and is fulfilled with

[13] Plato, *Sophistes,* 247 e; *The dialogues of Plato,* trans. Jowett (Random House, New York, 1937), vol. 2, p. 255.

[14] Plato, *Phaedrus,* 245 c 5, *ibid.,* vol. 1, p. 250; cf. J. B. Skemp, *The theory of motion in Plato's later dialogues* (Cambridge University Press, Cambridge, 1942), p. 3.

[15] Plato, *Timaeus,* 49 D; *The dialogues of Plato,* trans. Jowett (Random House, New York, 1937), vol. 2, p. 30.

them, and has become a visible animal containing the visible — the sensible God who is the image of the intellectual, the greatest, best, fairest, most perfect — the only-begotten heaven."[16] In this concluding passage of the *Timaeus* the notion of force is linked up again with the divine conception.

This metaphysical notion of force, however, is nowhere applied in Plato's writings when the explanation of an actual motion is needed. For instance, gravity as the cause of the falling motion of earthly objects has, in Plato's view, little to do with his general conception of force. Another unexplained assumption is introduced for this purpose: the tendency of like to join like. The common experience that "birds of a feather flock together" is taken over into natural philosophy and interpreted as an innate tendency of bodies of like nature to come together. Gravity becomes a quality, one is tempted to say, a chemical quality, without further rationalization. This maxim is given a quantitative aspect by stating that this tendency varies with the size of the object. It is not impossible that Plato's way of reasoning here was influenced by the atomists, for whom like atoms come together in vortices caused in turn by the unexplained eternal motion of the atoms. For the atomists, motion had to be an ultimate and unexplained fact, since all the contents of the universe was reduced to solid bodies without qualitative differences and nothing was left that could possibly be accounted for as the cause of motion. "For they say there is always movement. But why and what this movement is they do not say, nor, if the world moves in this way or that, do they tell us the cause of its doing so." [17]

For Plato, earth is thus attracted toward earth, water toward water, and fire toward fire.

If a person were to stand in that part of the universe which is the appointed place of fire, and where there is the great mass of fire to

[16] Plato, *Timaeus,* 92; *The dialogues of Plato,* trans. Jowett (Oxford University Press, London, 1924), vol. 3, p. 515.

[17] Aristotle, *Metaphysics,* lib. XII, cap. 6, 1071 b 33, in *The basic works of Aristotle,* ed. Richard McKeon (Random House, New York, 1941), p. 878.

which fiery bodies gather—if, I say, he were to ascend thither, and, having the power to do this, were to abstract particles of fire and put them in scales and weigh them, and then, raising the balance, were to draw the fire by force toward the uncongenial element of the air, it would be very evident that he could compel the smaller mass more readily than the larger; for when two things are simultaneously raised by one and the same power, the smaller body must necessarily yield to the superior power with less reluctance than the larger; and the larger body is called heavy and said to tend downwards and the smaller body is called light and said to tend upwards. And we may detect ourselves who are upon the earth doing precisely the same thing. For we often separate earthy natures, and sometimes earth itself, and draw them into the uncongenial element of air by force and contrary to nature, both clinging to their kindred elements.[18]

Heaviness and lightness, in other words, are explained by two assumptions: that like tends toward like, and that each element has its appointed region in space. The question may be asked whether one of these two assumptions would not be sufficient for the explanation of gravity. The exclusive adoption of the first principle, that is, that like joins like, would have introduced relativism which for the Greek mind would have been wholly uncongenial. The second alternative, that is, postulating merely appointed regions in space, or "natural places," for the elements, was, in fact, adopted by Aristotle. Through this geometrization of gravity, however, "the idea of gravitational attraction was for centuries expelled from natural philosophy." [19]

Furthermore, Aristotle's exclusive adoption of the principle of natural places led to the radical dichotomy of physical phenomena into celestial and terrestrial processes, each with its own autonomous laws of physics, a dichotomy that was to be discarded only with Newton's theory of universal attraction. In fact, Plato's axiom of "like joining like" would have been much more conducive to a unified, universal physical theory, especially

[18] Plato, *Timaeus,* 63 c; *The dialogues of Plato,* trans. Jowett (Random House, New York, 1937), vol. 2, pp. 42–43.

[19] A. E. Taylor, *A commentary to Plato's Timaeus* (Clarendon Press, Oxford, 1928), p. 441.

since Plato had alluded already to a uniform constitution of the world, for instance, in *Philebus:*

> We see that the elements which enter into nature of the bodies of all animals, fire, water, air, and, as the storm-tossed sailor cries, "land" (i.e. earth), reappear in the constitution of the world.[20]

How near ancient science could have come to the conception of a universal attraction can be seen from Simplicius' remarks in his commentary to Aristotle's *De caelo* where he says: "Others suggest as the cause, why celestial bodies do not fall towards the earth, some physical power, a centrifugal force [*periphoran*], which is greater than the original vertical inclination of fall [*rhope*] as stated by Empedocles and Anaxagoras."[21] Although Simplicius may have been wrong in attributing this idea to Empedocles or Anaxagoras, as has been pointed out by Gruppe,[22] the idea is certainly congenial to Plato's conceptual apparatus.

A similar allusion to some kind of universal gravity may be read into a passage in Plutarch's "Of the face appearing within the orb of the moon":

> The Moon has, for a help to preserve her from falling (on the earth), her motion and the impetuosity of her revolution. For every body is carried according to its natural motion, unless it be diverted by some other intervening cause. Wherefore the moon does not move according to the motion of her weight, her inclination being stopped and hindered by the violence of a circular motion.[23]

Although we may be right in going so far in our interpretation of these passages, assuming perhaps more intuition than purely logical inference implied, we should also remember that Plato's concept of force or power remains completely within the boundaries of his metaphysical doctrine and has hardly any

[20] Plato, *Philebus,* 29a, *The dialogues of Plato,* trans. Jowett, vol. 4, p. 597.

[21] Simplicius, *De caelo,* 91.

[22] O. F. Gruppe, *Die kosmischen Systeme der Griechen* (Reimer, Berlin, 1851), p. 194.

[23] Plutarch, *Moralia,* "Of the face appearing within the orb of the moon," ed. W. H. Goodwin (Little, Brown, Boston, 1871), p. 241.

bearing upon the problems of real physical situations. The term generally used by Plato to denote this idea of force is the word *dynamis.*[24] It is the noun corresponding to the verb *dynastai,* which means "to be able," "to be capable." The verb expresses not only the ability to act on something else but also the ability to be acted upon as a "patient." In contrast to our modern terms "force," "power," "activity," the Greek word *dynamis* signifies therefore not only transitive action or transeunt activity, but also passive susceptibility and receptibility. As "heat" can act on an object and make it warm, but can itself be reduced to coldness by another cold body, thus any other *dynamis* implies for the Greek mind both directions of activity. It is this double meaning that made *dynamis* a *terminus technicus,* especially in Greek medicine, as can be illustrated convincingly by examples from the Hippocratian treatise *On ancient medicine.*[25] As emphasized by Cornford,[26] Plato employed the term *dynamis* in this double meaning: he denoted the qualities of a substance, by which it makes itself sensible for us, by *dynamis,* an active transeunt process which Cornford aptly calls "the exteriorisation" of the substance; he also used *dynamis* for the characteristic susceptibility of the substance that leads to its individualization. The latter is the Platonic concept of force as expounded in the *Sophistes,* as mentioned above.

In order to understand fully the meaning of *dynamis* it is worth while to consider the correlative concept of *pathos,* derived from the Greek verb *paschein* ("to suffer"). *Pathos* denotes not only the passive suffering or sentimental emotion in a mind capable thereof, but also stands for any change of state or process and becomes nearly synonymous with "alteration" or

[24] Joseph Souilhé, *Étude sur le term "Dynamis" dans les dialogues de Platon* (Alcan, Paris, 1919).

[25] Hippocrates, *Ancient medicine,* in *Medical works,* trans. W. H. S. Jones (Loeb Classical Library; Harvard University Press, Cambridge, 1923), vol. 1, p. 37.

[26] Francis Cornford, *Plato's theory of knowledge* (Humanities Press, New York, 1951), p. 235.

"variation of qualification" (*alloiosis*). It is so used in Plato's *Hippias maior*,[27] in Aristotle's *Meteorology*,[28] in Simplicius' commentary on *De caelo*,[29] and most frequently, of course, in Galen's medical writings.[30] *Dynamis*, thus complementary to *pathos*, signifies power in the most general sense and retains this meaning throughout Greek scientific literature. Heat[31] and cold,[32] chemical activity,[33] botanical functions,[34] hardness,[35] and light[36] are spoken of as *dynameis*.

Galen[37] distinguishes not less than sixty different kinds of *dynamis* of the human body. Even earlier, for Aristotle the "strength" or "power" of the animal or of the human body was a *dynamis*.[38] It was from this context that Aristotle chose this term as a technical term for any kind of push or pull. Although Aristotle occasionally employs the term *ischys* for "strength" of the human body, it is the term *dynamis* that generally stands for the force of traction in Aristotelian "dynamics." To make the point quite clear: as we today speak of "dynamic personality" or "dynamic policy," using, so to say, a metaphor from physics, it is, strictly speaking, almost the reverse process that led Aristotle to adopt the term "dynamis," force, as a *terminus technicus* for his mechanics.

Actually, Aristotle recognizes two kinds of forces, the Platonic conception of force inherent in matter, which he calls "nature" (*physis*), and force as an emanation from substance, the force of

[27] Plato, *Hippias maior*, cap. VI, 285c; cf. also *Phaedon*, cap. 45, 96c.

[28] Aristotle, *Meteorology*, lib. IV, cap. 8, 385.

[29] Simplicius, *In librum de caelo*, I, 3, ed. Heiberg, p. 99.

[30] Galen, Symptomaton diaphoras, cap. 1, ed. Kuehn, vol. 7, p. 44.

[31] Aristotle, *Physics*, lib. 8, cap. 10, 266a.

[32] Aristotle, *Problemata*, sect. 7.

[33] Aristotle, *De sensu*, cap. 5, 443b; *Problemata*, cap. 23, 936a.

[34] Theophrastus, *De causis plantarum*, lib. 5, 1, 4, lib. 1, 15, 4.

[35] Theophrastus, *Fragmentum de lapidibus*, cap. 7, 40, ed. Wimmer (Leipzig, 1862), vol. 3, p. 43.

[36] Heron, *Pneumatica*, lib. 1, ed. Schmidt, p. 24.

[37] Galen, *Peri kraseōs kai dynameōs tōn haplōn pharmakōn*, lib. 2, cap. 2, ed. Kuehn, vol. 2, pp. 710ff.

[38] Aristotle, *Politica*, lib. VII, cap. 8, 1329a.

push and pull, causing the motion in a second object, and not in itself.

> Nature is a cause of movement in the thing itself, force a cause in something else, or in the thing itself regarded as something else. All movement is either natural or enforced, and force accelerates natural motion (e.g., that of a stone downwards), and is the sole cause of unnatural.[39]

The second type of force, stemming from substance, cannot, in Aristotle's view, be wholly detached from the substance in which it originated. This is the reason why force, as a technical concept in Aristotelian mechanics, is necessarily confined to only two forms of realization: push or pull. Since force is inseparable from its subject, it follows that the mover, as the subject of force, must be in constant contact with the motum, the object on which the force is exerted. All local motion must therefore be reducible to pull or push, as Aristotle explains in the seventh book of his *Physics:*

> It is clear, then, that in all cases of local movement there will be nothing between the mover and the moved, if it can be shown that the pushing or pulling agent must be in direct contact with the load. But this follows directly from our definitions, for pushing moves things away (either from the agent or from something else) to some other place, and pulling moves things from some other place either to the agent or to something else.[40]

For his mechanics proper, Aristotle confines himself solely to the concept of force as the agency involved in pulling or pushing, and ignores the Platonic concept of force as inherent in matter or what we may call today energy. Carteron's statement, "Entre la force comme principe de mouvement et la force comme quantité d'énergie correspondant à la puissance du mouvement en un

[39] Aristotle, *On the heavens,* 301 b 18 trans. W. K. C. Guthrie (Loeb Classical Library; Harvard University Press, Cambridge, 1939), p. 279.

[40] Aristotle, *Physics,* lib. VII, cap. 1, 243b, trans. P. H. Wicksteed and F. M. Cornford (Loeb Classical Library; Harvard University Press, Cambridge, 1934), vol. 2, p. 207.

moteur donné, la pensée d'Aristote oscille sans se fixer,"[41] may be correct generally speaking; but as soon as Aristotle inquires into the quantitative aspects of force he certainly fixes his ideas on the mechanical notion of force as exhibited in pull or push. Says he himself:

> Generally we speak of enforced action and necessity even in the case of inanimate things; for we say that a stone moves upwards and fire downwards on compulsion and by force; but when they move according to their natural internal tendency, we do not call the act one due to force; nor do we call it voluntary either; there is no name for this antithesis; but when they move contrary to this tendency, then we say they move by force.[42]

And a few lines later he says explicitly:

> For it is only when something external moves a thing, or brings it to rest against its own internal tendency, that we say this happens by force; otherwise we do not say that it happens by force.[43]

Even if we accept Spengel's opinion that the Eudemian Ethics, ascribed to Aristotle, is only a recast by some of his pupils, of his authentic Nicomachean Ethics, it certainly reflects Aristotle's own conceptions.

It is this notion of force as an agent of compulsory motion that Aristotle subjects to quantitative investigation and that forms the core of his mechanics. It is also this concept of force that corresponds most closely to the everyday experience of the Greek in those times. "Cette Dynamique, en effet, semble s'adapter si heureusement aux observations courantes qu'elle ne pouvait manquer de s'imposer, tout d'abord, à l'acceptation des premiers qui aient spéculé sur les forces et les mouvements."[44]

[41] Henri Carteron, *La notion de force dans le système d'Aristote* (Vrin, Paris, 1924), p. 22.

[42] Aristotle, *Ethica Eudemia de virtutibus et vitiis,* lib. 2, cap. 8, 1224 a 15, *The works of Aristotle,* trans. under the editorship of W. D. Ross (Oxford University Press, London, 1940), vol. 9.

[43] *Ibid.,* 1224 b 6.

[44] Pierre Duhem, *Le système du monde* (Hermann, Paris, 1913), vol. 1, p. 194.

In the seventh book of his *Physics* Aristotle outlines his laws of forces as follows:

That which is causing motion is always moving something in something and up to somewhere. (By the something "in which" I mean time, and by the "up to somewhere" the measure of the distance traversed; for if a thing is now causing motion, it has already caused motion before now, so that there is always a distance that has been covered and a time that has been taken.) If, then, A is the moving agent, B the mobile, C the distance traversed and D the time taken, then A will move $\frac{1}{2}B$ over the distance $2C$ in time D and A will move $\frac{1}{2}B$ over the distance C in time $\frac{1}{2}D$, for so the proportion will be observed. Again, if A will move B over distance C in time D and A will move B over distance $\frac{1}{2}C$ in time $\frac{1}{2}D$, then E $(=\frac{1}{2}A)$ will move F $(=\frac{1}{2}B)$ over the distance C in time D; for the relation of the force E $(\frac{1}{2}A)$ to the load F $(\frac{1}{2}B)$ in the last proposition is the same as the relation of the force A to the load B in the first, and accordingly the same distance (C) will be covered in the same time (D). Also if E $(\frac{1}{2}A)$ will move F $(\frac{1}{2}B)$ over distance C in time D, it follows that in the same time E will move 2 F's over half the distance C. But if A will move B over the whole distance C in time D, half A (E) will not be able to move B, in time D or in any fraction of it, over a part of C bearing the same proportion to the whole of C that E bears to A. Because it may well happen that E cannot move B at all; for it does not follow that if the whole force could move it so far, half the force could move it either any particular distance or in any time whatever; for if it were so, then a single man could haul the ship through a distance whose ratio to the whole distance is equal to the ratio of his individual force to the whole force of the gang.[45]

Nowhere in this context does Aristotle even allude to the units in which he conceives this force A to be measured. It may be suggested that a unit of weight might have served for the measurement of force, as weight is most certainly the earliest type of force to be measured and is the one force in terms of which, until modern times, all other forces were expressed. Undoubtedly, when the Greek wanted to know with what force he was working on the lever of his oil press he could measure it in terms of weight. For Aristotle, however, weight was the manifestation of natural motion and not the cause of compulsory; in other words,

[45] Aristotle, *Physics,* lib. VII, cap. 5, 249 b 26–250 a 20, trans. P. H. Wicksteed and F. M. Cornford (Loeb Classical Library; Harvard University Press, Cambridge, 1934), vol. 2, pp. 257–259.

it was not a force of push or pull and, consequently, could not be employed as the standard for measurement of forces. The absence of a suitable and reproducible standard of force, in the Aristotelian sense of the concept, or, in other words, the omission of elastic forces from scientific consideration, led Peripatetic physics in its treatment of forces to lose contact with immediate experience and experimentation, which betrays itself as late as in the treatises on forces by the *calculatores* of the fourteenth century.[46] From the concluding remarks of the long quotation given from the seventh book of Aristotle's *Physics* we also note that friction is not yet conceived as a kind of force, although Aristotle includes an external agent that "brings a body to rest against its own internal tendency"[47] under the category of forces. However, since he is not yet in possession of a principle of inertia, the continuation of a compulsory motion cannot be counted as an "internal tendency," and friction, therefore, cannot be recognized as a force. As with the moving force A, so also with the mobile B, it is not clear at all in what units this latter should be measured; it certainly would be an anachronism to interpret B as mass or weight, let alone inertia, since ancient Greek science clearly had not even clear conceptions of these notions.

In the case of natural motion, that is, downward motion of heavy objects or upward motion of light objects, Aristotle discusses the resistance of the medium in detail[48] and infers that the velocity of the mobile is given by the ratio (*analogia,* 215 b 3) of the corresponding densities:

$$v = \frac{A}{B},$$

where A, the motive power or motive force, is determined by the density of the mobile and B, the resistance, by the density of the

[46] See, for example, Thomas Bradwardine, *Tractatus de proportionibus* (1328), ed. and trans. H. L. Crosby, Jr. (University of Wisconsin Press, Madison, 1955).

[47] Aristotle, *Ethica Eudemia de virtutibus et vitiis,* 1224 b 6.

[48] Aristotle, *Physics,* 215 b 1–216 a 8 (Loeb edition), vol. 1, pp. 350–355.

medium. This law of motion, thus interpreted, was accepted by the majority of the Aristotelians and in particular by Averroes in his commentary on what was called in the middle ages "Text 71" (*Physics,* 215 a–215 b 20). On the other hand, it became also the object of severe criticism by Joannes Philoponus, by Avempace and, eventually, by Galileo Galilei. For a detailed history of these criticisms the reader is referred to Ernest A. Moody's important article, "Galileo and Avempace."[49] Aristotle's law of enforced motion, if expressed in modern symbols, is essentially identical with the law just mentioned. If, as before, A signifies the moving force and B the resistance, and if C is the distance traversed and D the time involved, Aristotle's lengthy discussion can be summarized by the following formula:

$$\frac{A \cdot D}{B \cdot C} = \text{constant.}$$

Hence A/B is proportional to C/D, which is the velocity of the mobile.

Since force as an action at a distance has no place in Aristotle's conceptual scheme, an explanation of the motion of planets and stars in the heavens can be furnished only by either supposing an external agent, a prime mover, an astral intelligence, or endowing the stars with a life of their own. Both explanations seemed for Aristotle to be equally acceptable, although the status of astral intelligences in Aristotle's teaching remains most obscure. In his *Metaphysics*[50] he alludes to the conception of an active astral life of each planetary sphere. In explaining the regularity of celestial motions, however, Aristotle refers to the prime mover:

> Everything moved is moved by something, hence the irregularity of the movement must proceed either from the mover or from the object moved or from both. If the mover does not act with constant force, or if the object changes instead of remaining constant, or if both alter, then there is nothing to prevent the movement of the object from being ir-

[49] E. A. Moody, "Galileo and Avempace," *Journal of the History of Ideas 12,* 163–193, 375–422 (1951).

[50] Aristotle, *Metaphysics,* book XII, chapter 8, 1073 a 28ff.

regular. But none of these hypotheses can be applied to the heaven; for the object of the movement has been demonstrated to be primary, simple, ungenerated, indestructible, and altogether changeless, and we may take it that the mover has far better reason to be so: only what is primary can move the primary, what is simple the simple, what is indestructible and ungenerated the indestructible and ungenerated.[51]

No further important development of the concept of force took place during the time of the early Lyceum or the early Academy. An unsuccessful yet notable attempt was made by Pappus, the greatest mathematician of the third century, to bring gravity within the scope of the Aristotelian conception of pulling force. Pappus imagines a body of given weight placed on a horizontal plane and assumes that a certain force (*dynamis*) is necessary in order to move this body. He now inquires into the magnitude of this force, if the plane is inclined to the horizontal at a given angle.[52]

Even Archimedes, the founder of statics, has little to contribute to the development of the concept of force. His treatment of mechanics is a purely geometric one, implying the notions of distance and weight; but weight is assumed as a given, primitive notion without any further analysis.

It is only with the Stoa that the history of the concept of force shows a new turn. The problem that initiates this new development is the query for an explanation of the connection between the tides and the movement of the sun and moon. The problem is an old one. Already Aristotle, Dicaearchos, and Pytheas had tried to solve it. Aristotle, for whom the idea of an action at a distance was an impossibility, had to stress his scheme of pulling and pushing forces to furnish an explanation: the sun, he supposes, moves the winds and these, falling upon the Atlantic Ocean, push,

[51] Aristotle, *On the heavens,* trans. W. K. C. Guthrie (Loeb Classical Library; Harvard University Press, Cambridge, 1939), pp. 171–172.

[52] Pappus, *Opera,* lib. VIII, prop. 9 (ed. Hultsch, Berlin, 1876), vol. 3, p. 1055. Cf. Pappus d'Alexandrie, *La collection mathématique,* trans. Paul Ver Eecke (Desclée de Brouwer, Paris and Bruges, 1933), vol. 2, p. 833. The problem reads: "Having given the force which draws a given weight along a horizontal plane, find the additional force which will draw the same weight along a given inclined plane."

in their turn, the waters of the sea. In the *Timaeus,* the great rivers that flow into the ocean are held to be the cause of the tides. A similar explanation is given by Crates of Mallos, who deduced the tides from certain currents of oceanic waters. Seleukos, who believed in the rotational motion of the earth, accounts for the tides by the motion of the moon which moves in the opposite direction to the rotation of the earth and acts on the intermediate air which in its turn exerts a pressure on the water, thereby raising undulatory motions.[53]

A completely different approach to this problem is chosen by Poseidonius. He traveled to Gades and stayed there for thirty days to study the phenomenon of the tides. Like a modern scientist he tried first to collect all the relevant facts and data. No doubt, the tides showed a periodicity dependent on the periodicity of the celestial bodies. Like the stars, the ocean exhibited a diurnal, monthly, and annual period; the height of the flow was greatest during the summer solstice, he was told by the neighboring population. In search for an explanation of this majestic phenomenon, Poseidonius interpreted the doctrine of *tonos,* of a universal tension, formulated by his predecessor Chrysippus, in a dynamic manner. The Chrysippian theory of mutual connections and tensions was originally based on an assumed materialization of Aristotelian concepts and forms.

For Poseidonius the phenomenon of tides is a manifestation of forces pervading all space, but forces completely at variance with the Aristotelian conception. With Aristotle, force, although the emanation of substance, had its location in the subject, the carrier of the force, and was wholly independent, at first, of the object on which it was exerted by contact. Poseidonius, on the other hand, who posits force as the primary and most fundamental notion in his natural philosophy, conceives force as an expression linking, and simultaneous with, the two objects related thereby. Force becomes a mutual correspondence of action, a

[53] Aetius, III, 17. Cf. *Aetii Amideni libri XVI,* Latin trans. by Cornarius (Basel, 1542).

"sympathy" in the original sense of the word. The universe becomes one single whole by the interaction of a system of forces. "So steht Poseidonius selbständig als Philosoph und Systematiker und ebenbürtig neben Aristoteles und Chrysipp. Der erste lehrt: die Welt ist zu erklären aus dem Begriff; der zweite lehrt: die Welt ist zu erklären aus der Vernunft. Der dritte lehrt: die Welt ist zu erklären aus der Kraft." [54]

The Stoic conception of force, according to which the agent and the patient are inseparable in the formation of the connecting activity, is closely related to their doctrine of the unity of the cosmos. The ancient Greek idea of "sympathy" in medicine and the recognition that one part of the body is affected by the disease of another part, an idea found already in the Hippocratic writings and described in detail later by Galen,[55] is thus transferred by the Stoa to the cosmos as a whole. The consonant vibration of one string of the harp when another string is struck[56] or the yawning of one person evoked by the yawning of another and many similar examples are taken as indications demonstrating the internal interconnection of the parts of the world. According to Alexander of Aphrodisias, "sympathy" (*sympatheia*) acts by means of the *pneuma*,[57] an all-pervading fluid. In the Neo-Platonic school of thought this Stoic *pneuma* is interpreted as a

[54] Karl Reinhardt, *Poseidonius* (Beck, Munich, 1921), p. 11. See also Karl Reinhardt, *Kosmos und Sympathie* (Beck, Munich, 1926).

[55] "For those people who do not believe that there exists in any part of the animal a faculty for attracting its own special quality ["attractricem convenientis qualitatis vim" (Linacre), *helktikēn dynamin* in Greek] are compelled repeatedly to deny obvious facts." Galen, *On the natural faculties,* trans. A. J. Brock (Loeb Classical Library; Harvard University Press, Cambridge, 1928), p. 49. In his discussion of the contrast between the vitalists and the Epicurean atomists, Galen explains the doctrine according to which Nature "puts together the bodies both of plants and animals; and this she does by virtue of certain faculties, which she possesses — these being, on the one hand, attractive and assimilative of what is appropriate, and on the other, expulsive of what is foreign." *Ibid.,* p. 45. The conception of an attractive force (*helktikē dynamis*), inherent in Nature, plays an important role in Galen's writings.

[56] Asclepius of Tralles, *In Aristotelis Metaphysicorum libros commentaria,* ed. Michael Haydruck (Berlin, 1888), vol. 7, 2, p. 92.

[57] Alexander of Aphrodisias, *Peri kraseōs kai auxeseōs, Scripta minora,* ed. Ivo Bruns (Berlin, 1892), part 2, p. 216.

soul (*psyche*) inherent in the world, with reference to the Platonic idea of a world soul.[58]

The introduction of a substratum, a material or somehow spiritual fluid, assumed as necessary for the *modus operandi* of the "sympathy," is a concession to the Aristotelian conception of force. For since the days of Empedocles[59] and throughout Aristotelian physics, as we have seen, force, conceived as *"dynamis,"* was held as a certain fluid or emanation of substance, while Poseidonius' conception of force as a "sympathy" came very near to the notion of action at a distance as conceived by classical physics of the seventeenth and eighteenth centuries, since it posited the appearance of one phenomenon *ex opere operato* by the existence of another, without inquiring into the causal connexus at all.

The force conception of "sympathy" plays an important role in the scientific, philosophical, and mystical literature of the Roman period. Plotinus, in his *Enneads*,[60] reduces action at a distance repeatedly to "sympathy." In Cicero's writings the concept of "sympathy" is frequently mentioned under the names of *coniunctio, contagio, cognatio, continuatio, convenientia, concentus* and *consensus rerum*.[61]

Magic and astrology, two important subjects in the intellectual life of the Roman period and the Middle Ages, seem to have found their rational justification in Plotinus' conception of force. As he explains in detail,[62] magic is possible through the sympathy of the like and the contrast of the unlike. Since sympathy is the excitation of a "pathos" in one object through the "pathos"

[58] Proclus, *In Platonis Timaeum commentaria*, ed. E. Diehl (Leipzig, 1903–1906), vol. 1, p. 412.

[59] See p. 27.

[60] "This results from the interrelating community existing within the universe." Plotinus, *Enneades*, II, 1, 7. Cf. also Plotinus, "The stars know our desires through the agreement and sympathy established between them and us," *Complete works*, trans. K. S. Guthrie (Bell, London, 1918), vol. 2, p. 477.

[61] Cicero, *De divinitate*, lib. II, cap. 14, 60, 69.

[62] "Likewise, magic is founded on the harmony of the universe; it acts by means of the forces which are interconnected by sympathy." Plotinus, *Complete works*, p. 478.

in another,[63] sympathetic magic is the stimulation of a certain state in one part of the universe and the belief that a corresponding state is produced elsewhere.

Ironically, it was to Democritus that a whole literature of sympathetic magic was ascribed at this period — the so-called Pseudo-Democritian writings, which were so instrumental for the spread of superstition in the late Roman and early medieval times. Pliny's *Natural history*[64] is based extensively on this literature of *mirabilia.* Pliny himself asserts repeatedly in his treatise[65] that its purpose is the clarification of "sympathy," which he frequently calls *concordia,* and the exposition of its various forms.

Pliny's uncritical use of *vis,*[66] corresponding to the Greek *dynamis,* for the designation of physical, chemical, medical, and occult forces is partly responsible for the obscurity of his writings, which appear as a peculiar mixture of natural science and superstition. Occult powers were indiscriminately regarded as physical forces, and physical force was considered as magic in its essence. An interesting example that illustrates this intermixture of conceptions is Pliny's reference to the echeneïs, a little fish (later called "remora") which was supposed to be able to stop moving ships by simply attaching itself to their keels. This supposedly magical power is interpreted by Pliny as a serious physical force, made comprehensible only by the conception of sympathy and antipathy. In book 32 of his *Natural history* Pliny relates:

> And yet all these forces, though acting in unison, and impelling in the same direction, a single fish, and that of a very diminutive size — the fish known as the "echeneis" — possesses the power of counteracting.
>
> Winds may blow and storms may rage, and yet the echeneis controls their fury, restrains their mighty force, and bids ships stand still in their

[63] Cf. Excerpta ex Theodoto: "Hē gar sympatheia pathos tinos dia pathous heterou," Clemens Alexandrinus, ed. Stählin (Leipzig, 1905), vol. 3, p. 117.

[64] Caius Plinius Secundus, *Naturalis historia. The Natural history of Pliny,* trans. John Bostock and H. T. Riley (Bohn, London, 1856).

[65] *Ibid.,* beginning of book XX; book XXXVII, 59.

[66] *Ibid.,* book XXIII, 3; XXIV, 3; XXV, 80 (vol. 4, pp. 2, 56, 158).

career; a result which no cables, no anchors, from their ponderousness quite incapable of being weighed, could ever have produced! A fish bridles the impetuous violence of the deep, and subdues the frantic rage of the universe — and all this by no effort of its own, no act of resistance on its part, no act at all, in fact, but that of adhering to the bark! Trifling as this object would appear, it suffices to counteract all these forces combined, and to forbid the ship to pass onward in its way! [67]

Pliny even supplies documentary evidence, mentioning the battle of Actium, in which this little fish stopped the mighty praetorian ship of Antonius, and the arrest of Caius Caligula's ship when returning from Astura to Antium. With reference to Caligula, Pliny says: "One thing, too, it is well known, more particularly surprised him, how it was possible that the fish, while adhering to the ship, should arrest its progress, and yet should have no such power when brought on board." [68] Even for Pliny something in this story seems to be "fishy!"

In later Latin literature, too, the concepts of physical force and occult agencies are indiscriminately intermingled. The term *virtus*, in general, denotes them both. But worse, since the Vulgate usually translated the Greek *dynamis* with *virtus*, the very same term was also employed for the denotation of forces in the religious sense. Thus, Isidore of Seville speaks of the *virtus* of the sacraments: "Ob id Sacramenta dicuntur, quia sub tegumento corporalium rerum virtus divina secretius salutem eorundem sacramentorum operatur, unde et a secretis virtutibus, vel a sacris sacramenta dicuntur." [69]

The concept of "sympathy" was also employed in order to give astrology its logical support. Through "sympathy" Sextus Empiricus[70] explains the influence of the moon on the tides, as did Plotinus, and also its influence on living beings. Proclus, in generalizing this idea, asserts the influence of the celestial spheres on

[67] *Ibid.*, vol. 6, p. 2.

[68] *Ibid.*, p. 3. Cf. Lucan, *Pharsalia*, book VI, 670.

[69] Isidore of Seville, *Etymologiarum opus*, lib. VI, cap. 40, in J. P. Migne, *Patrologiae cursus completus*, vol. 82 (Paris, 1850), column 255. Cf. *Corpus grammaticorum Latinorum Veterum*, ed. F. Lindemann (Leipzig, 1831), vol. 3.

[70] Sextus Empiricus, *Pros physikōs*, lib. I, cap. 78, ed. Becker (Berlin, 1842), p. 409.

mundane events. Plotinus, in the second book of his *Enneades*, raises the question: "Whether the stars effect anything?" For him stars are, as they were for Plato, eternal and divine living beings, as expression of the omniscience of the world soul. In opposition to the faithful Aristotelians, who like Ptolemy[71] in his *Tetrabiblos* rejected all actions at a distance and tried to explain the tides by forces of contact (atmospheric currents, heat, and light), Plotinus reduced cosmic and astrological influences to the relation of "sympathy" that exists, in his view, between all parts of the universe. From the *Liber de mysteriis*[72] Porphyry and Iamblichus may be quoted as similar exponents of this conception of force.

Before we proceed to expound the conception of force in medieval astrology, we have to discuss, because of its importance for the subsequent history of the concept of force, the spiritual and religious explication of the concept of force in the Jewish-Alexandrian school of thought. Let us examine, therefore, the concept of force as expounded by Philo, the most important representative of this school of thought. His concepts are naturally biased by his deeply religious outlook but are also, in part, at least, derived from his anti-Stoicism.

In analyzing Philo's conceptions, the religious and somehow mystical climate of opinion in his time has to be taken into consideration. Stars and angels, whose existence was not doubted at that time, were recognized as manifestations of divine powers. In fact, the Septuagint, in translating Psalm 24, verses 9–10, "Open wide your gates, ye chiefs . . . that the king of glory may come in. Who is the king of glory? The Lord of Zebaoth. He is the king of glory," translated the expression "Lord of Zebaoth"[73] as "Lord of forces" (*Kyrios tōn dynameōn*), demonstrating the dogma, held generally at that time, that all forces in nature are

[71] Proclus, *In rem publicam*, ed. W. Kroll (Leipzig, 1899–1901), vol. 2, p. 258.

[72] *Liber de mysteriis* (Aldine Press, Venice, 1497, 1516). See also Carl Rasche, *De Iamblicho libri qui inscribitur de mysteriis auctore* (Aschendorff, 1911).

[73] "Zebaoth" in Hebrew stands for "hosts," "armies," etc. Cf. Isa. 13:4, "The Lord of hosts mustereth the host of the battle."

but manifestations of divine power. The same idea is found in Aristobulus, who is often regarded as the founder of the Jewish-Alexandrian school — "The power of God is through all things" [74] — and is even corroborated in the Letter of Aristeas, dating from the time of Ptolemy Philadelphus, in which Elazar is reported to have said: "There is only one God . . . His power [*dynamis*] is manifested throughout the universe, since every place is filled with His might." [75] In conformity with these ideas, Philo interprets Deuteronomy 4:39, "God in heaven above and upon the earth beneath," as meaning not God Himself, but his forces in the sky and on the earth. Philo says: "He has made His forces extend through earth and water, air and heaven, and left no part of the universe destitute, and by uniting all with all has bound them fast with invisible bonds." [76] Philo thus postulates the existence of invisible bonds of forces throughout the universe, as did the Stoa. But Stoic thought, even of its most religious display, seems unacceptable to him; for Stoic religious thought, employing the concept of *pneuma,* converged to the principle of the immanence of material forces and thereby to the immanence of God in the universe. For Philo the God of the Bible remained a Personal God, although not personified. He rejected, therefore, the Stoic conception of force as implying that the forces of God are immanent only in the world and precluding the existence of an incorporeal being outside the world. The forces of the Stoic conception seem to him to be active by their own nature, whereas he derives their activity from some power above them. "To act is the property of God, and this we may not assign to any created thing; the property of the created is to suffer." [77] However, when we ascribe forces to created beings, says Philo, we refer to the

[74] Cf. Eusebius, *Preparatio evangelica,* VIII, 12, 666d.

[75] *De Bibliorum textibus originalibus, versionibus Graecis et Latina Vulgata,* ed. Hody (Oxford, 1705).

[76] Philo Judaeus, *De confusione linguarum,* 27, 136. Cf. H. A. Wolfson, *Philo* (Harvard University Press, Cambridge, 1948), vol. 1, p. 344.

[77] Philo Judaeus, *De cherubim,* 24, 77, in *Philonis Alexandrini opera,* ed. L. Cohn and P. Wendland, vol. 1 (Reimer, Berlin, 1896).

power of God invested in them and acting indirectly through their intermediaries. God bestows many of his salutary forces "through earth, water, air, sun, moon, and heaven."[78] Thus, he believes, he can account for the permanency of shapes and of qualities in the things of the world. Force, therefore, according to Philo, exists in three variations: as the eternal property of God, identical with his essence; as incorporeal beings created by God prior to the creation of the world; and as wholly immanent in the world. Angels, in his view, are but a special form of such immanent forces in the world. Through these forces Philo, in conformity with Platonic conceptions (see page 34), explains the reality of natural objects, and the cause of their differentiation from otherwise indistinguishable matter. "These powers, so far as they affect the material world, correspond with what we designate as physical forces."[79]

The notion of "mind" (*nous*) as invented by the Greeks, combined with the Platonic conception of the "idea" and the Neo-Platonic notion of "logos," was hypostatized to a spiritual substance, residing beyond the highest sphere of the universe, separated from its counterpole, the earth, by concentric spheres of ever-increasing materiality. Forces and beings of dual nature, half corporeal and half spiritual, are presupposed as the intermediaries between the two extremes, God and the earth. Alexandrian Judaism and early Christianity transform this dynamic-geometric representation into an ethical interpretation. Infinite, divine force (*dynamis*) thus becomes the highest and foremost principle, both in degree and in time, for the conceptual system of Simon Magus, who introduced the conception of forces (*dynameis*) into the teachings of the Gnostic schools. His conception of force is a generalization of the notion of *dynamis* referred to in Luke 6:19 and 8:46 as well as in Mark 5:30. The concept of force, which for the Peripatetics has already lost much of the

[78] Philo Judaeus, *De mutatione nominum,* 8 (I, 587), *ibid.,* vol. 3 (1898).

[79] James Drummond, *Philo Judaeus* (Williams and Norgate, London, 1888), vol. 2, p. 69.

mythological and magical content that it had possessed in pre-Aristotelian thought, is endowed again with an autonomous existence and reinstalled as a supernatural agency of divine design. It is this conception of force that is taken up again by the religious philosophers of the Middle Ages and employed, for example, in their explanation of celestial motions. Philo's conception of force can also be regarded as a possible source of the Cabbalistic teachings of emanating forces. Judah ben Samuel Halevi, the famous Hebrew poet of the early twelfth century, expresses certain similar ideas when he explains the etymology of the Hebrew word "Elohim," the appellation of God. Prior to Judaism, he maintains, every process of motion in nature was conceived as enacted by a separate divinity; the plural form of God's name, he claims, is reminiscent of that time, until with the advent of Judaism all forces were held to have their origin and fount in the Only One.[80] Similar explanations were set forth by Moses Maimonides, for whom all forces are angels, and by Abraham ibn Daud, who derives all forces of nature from God, with reference to Exodus 15:11.

Based on the Platonic idea of a world soul, immanent in nature, on the Aristotelian doctrine of motion applied to celestial bodies, on the Neo-Platonic conception of an all-pervading force, and finally on Philo's religious interpretation of cosmic forces, the belief that divine intelligences or angels are the motive forces behind celestial motion becomes a fundamental assumption during the Middle Ages, shared by almost all ecclesiastical authorities. From the early Enoch literature, which characteristically placed the archangel Gabriel[81] over all powers and which described the revolutions of the stars "according to the number of the angels,"[82] through John Damascene's conception of an intelli-

[80] David Kaufmann, *Geschichte der Attributenlehre in der jüdischen Religionsphilosophie des Mittelalters* (Perthes, Gotha, 1877), p. 109.

[81] Gabriel is etymologically derived from *gever* ("man") and *El* ("God"); *gever* is related to *g'vurah* (see p. 21), as the Latin *vir* is to *vis*(?).

[82] Lynn Thorndike, *A history of magic and experimental science* (Macmillan, New York, 1923), vol. 1, p. 343.

gence presiding over the celestial order, until Thomas Aquinas' *Summa theologica*[83] and even until the astrological writings of Johannes Kepler, celestial motion was conceived as an incontestable manifestation of divine agencies. Thomas Aquinas' reasoning in his *De substantiis separatis*[84] is characteristic of the trend of ideas involved. With reference to the *Physics* (256a–267 b) of Aristotle, whose conceptions seem to Aquinas "clearer and more sound" than those of any other approach, it is contended that no motion occurs without a mover; to avoid an infinite regress of movers, a first mover must be assumed who is himself unmoved. Since the motion under discussion is an eternal motion, the force, as the cause of this motion, must be an infinite force, and consequently cannot be originated by a corporeal being that possesses only finite qualities. But since only bodies can be moved, the first moved must be a body animated by an eternal intellect; and this is the first heaven.

It would lead us too far astray to discuss Aquinas' identification of motion with the intellectual desire of the good-in-itself, a series of syllogistic conclusions by which he attempts to correlate kinematic or dynamic concepts with ethical notions. What is more important for our subject is the question, much discussed in the Middle Ages, how the union of the celestial body with the intelligence is established. For it is generally held that the celestial bodies can be considered as animated only in the sense of a virtual contact of the intellectual mover with the moved object. The intelligences are not united with the body as soul is united formally to a body in discharging its numerous functions, or, to use an example mentioned much later by Giacomo Zabarella in his *Commentaries on De anima,* the intelligence is not inherent in the celestial body as is the shape of a ship in the ship itself — a ship without the shape of a ship being unimaginable — but the relation of the intelligence to the celestial body is rather like

[83] St. Thomas Aquinas, *Summa theologica,* lib. I, q. 70 et sequ.

[84] St. Thomas Aquinas, *De substantiis separatis, Opuscula omnia* (Lethielleux, Paris, 1927), vol. 1.

that of the pilot to his ship. According to Aquinas — and this view is again characteristic of the majority of ecclesiastical authorities — the intelligences are not *intelligentiae informantes,* but *intelligentiae assistentes* performing the duty of an external *spiritus rector;* yet their union with the body is nevertheless a dynamic union. But now the question arises, how then could the motion of a star be regarded, in the Aristotelian sense[85] of the word, as a natural motion? Aquinas reconciles Aristotelian conceptions with the religious outlook by stating that celestial motion is natural not by virtue of its active principle, but by reason of the "aptitude that the passive principle has to be moved by the intellect through virtual contact." The motion is natural, because it is the nature of the celestial body not to impede this kind of movement.

Aquinas' doctrine of celestial intelligences and his thesis that "all multitude proceeds from unity" lead him to accept the principles of dynamic astrology. Since terrestrial motions, being not circular but in a straight line, have to depend on the perfect motions of the heavens, celestial bodies are the causes of terrestrial motions. "Therefore the motions of these inferior bodies, being various and manifold, are related to the motion of the celestial bodies as to their cause." [86]

[85] Aristotle, *De caelo,* 269 b 3; *Physics,* 255 a 6.

[86] St. Thomas Aquinas, *Summa theologica,* I a, q. 115, a 3.

CHAPTER 4

CONCEPTS OF FORCE IN PRECLASSICAL MECHANICS

The belief in the divine nature, or in the divine origin, of the celestial motive powers is characteristic of ancient and medieval science. Pierre Duhem dates the beginning of modern science from the moment when science dispensed with this assumption:

> If one wishes to draw a line of separation between the realm of ancient and modern science, it must be drawn at the instant when Jean Buridan conceived his theory of momentum, when he gave up the idea that stars are kept in motion by certain divine intelligences, and when he proclaimed that both celestial and earthly motions are subject to the same mechanical laws.[1]

One could, however, equally well draw the line of demarcation at a much later date, when Newton published his law of universal attraction and promoted thereby the principle of a universal science which does not discriminate between superior, the celestial, and inferior, the terrestrial. While Buridan was one of the first who revolted against the Neo-Platonic conception of force as a divine being, Newton was the one to give the *coup de grâce* to the doctrine of astrological forces. For Newton's universal law of gravitation supplanted another and different universal natural law which had a long reign in man's intellectual life, a universal

[1] Pierre Duhem, *Études sur Léonard de Vinci* (De Nobele, Paris, 1955), vol. 3, p. ix.

law that operated also with forces: astrology, dealing with celestial influences exerted by the superior stars upon the inferior mundane world. As Lynn Thorndike has pointed out recently in a concise and noteworthy article,[2] the importance of astrology as a consistent edifice of thought for the Middle Ages cannot be overestimated for a correct appraisal of pre-Newtonian science.

The importance of astrological speculations, in particular, for the history of the concept of force can readily be understood when we read, for example, what George Sarton has to say about Roger Bacon:

> However, Bacon's thoughts did not dwell so much on statics as on dynamics. He was pondering on the nature of force, especially on force or action at a distance. Curiously enough, these thoughts, earnest as they were, were partly astrological. For among the forces or actions considered by him were light and gravity, but also astrological influences, the reality of which were beyond doubt. How were these astrological influences transmitted across the open spaces? How were these distant causalities propagated? It was very remarkable to ask such questions, and we must not blame him for failing to solve them.[3]

When we speak of astrology we do not imply, of course, solely genethlialogy or the prognostication by horoscope for the individual from his hour of nativity. In our present context we use the term "astrology" to mean the general assumption that the entire world of nature is ruled and determined by the movements of the celestial bodies. This assumption seems to be based on an intuitive recognition that causality is associated with certain distributions in space, an idea which, as we shall see later on, underlies the modern concept of force as well. It is, however, the peculiar conception of force in astrology, the "celestial virtue," that introduces the irrational element into the subject and degrades it to a pseudo science by making it inaccessible to mathematical demonstration and experimental verification. In this profound

[2] Lynn Thorndike, "The true place of astrology in the history of science," *Isis 46*, 273 (1955).

[3] George Sarton, *Introduction to the history of science* (Williams and Wilkins, Baltimore, 1931), vol. 2, part 2, p. 763.

way Cassirer analyzes the contrast between astrology in our sense and modern science:

Die moderne Physik "erklärt" alles räumliche Beisammensein, alle Koexistenz der Dinge zuletzt damit, dass sie sie auf Bewegungesformen und Bewegungsgesetze zurückführt. Ihr wird der physische Raum zum Kraftraum, der sich aus dem Ineinandergreifen der "Kraftlinien" aufbaut. Ihre letzte und deutlichste Ausprägung hat diese Grundansicht in der allgemeinen Relativitätstheorie erhalten, in der die Begriffe des metrischen Feldes und des Kraftfeldes ineinander übergehen, in der das Dynamische metrisch, aber auch eben so sehr das Metrische dynamisch bestimmt wird. Wenn hier der Raum in Kraft, so wird dagegen im astrologischen Denken die Kraft in Raum aufgeloest. Das Himmelsgebäude und die Stellung und Gliederung seiner einzelnen Teile ist selbst nichts anderes als die Anschauung des Wirkungszusammenhanges des Universums.[4]

It is, of course, beyond the scope of this study to give a detailed analysis of the various conceptions of "celestial virtues" throughout the long history of astrology. Let us, therefore, again select only one representative example, which moreover has special importance for our subject.

Astrology was introduced into the Muslim world during the eighth century, when the Caliph Al-Mansur, with the assistance of the learned Jew Jacob ben Tarik, founded in Bagdad an academy for mathematical sciences. Here, among such other noted Arabian astrologers as Messahala, Albategnius, Alpetragius, Alchabitius, and Abenragel, Abu Mashar Ja'far ibn Muhammad ibn Umar al-Balkhi, or, in his Latin name, Albumasar, received his training. In his great *Book of introduction to astronomy* (*Kitāb al-mudkhal ilā 'ilm ahkam al-nujūm*)[5] he defends astrology on the grounds that astral bodies do not consist of one of the traditional elements of this world, nor of a combination of them. In contrast to the growth and decay inherent in the sublunary sphere, the motion of the heavenly bodies is eternal, per-

[4] Ernst Cassirer, *Die Begriffsform im mythischen Denken* (Teubner, Leipzig, 1922), p. 43.

[5] Albumasar's treatise was known in Europe under the title *Introductorium in astronomiam Albumasaris Abalachi, octo continens libros partiales* and was printed in Augsburg in 1489 (copy in the New York Public Library), and later also in Venice (1495, 1506).

fect, and changeless. However, corresponding to their circular motions and periodicities, cyclic motions of growth and decay take place also in the sublunary world, caused by the eternal circular motion of the heavenly bodies. Thus, in his view, the celestial essence exerts a recognizable influence on everything mundane and thereby the inferior world is bound up with the superior. Albumasar tried to show, in particular, how the tides are caused by astrological influences, without, however, specifying the kind and character of influences he has in mind. Although he knew, for instance, that the light of the moon is irrelevant for the explanation of the tides, since the tides are produced under the full moon as well as under the new moon, yet he made no positive statement regarding the physical nature of the influences or forces involved. Albumasar's work was translated later by John of Seville (Joannis Hispalensis) and by Herman of Dalmatia (Hermannus Secundus-Dalmata) and these translations when carried north from Spain transmitted astrological knowledge to the Christian schools of Europe.

Albumasar's treatise illustrates clearly how Arabian astrology was founded on Aristotelian conceptions; indeed, most of Muslim science was merely a recapitulation or elaboration of the teachings of the Stagyrite, optics, alchemy, and certain branches of medicine perhaps excepted. In particular, if we ignore for the moment certain early Arabian proponents of the impetus theory, Aristotelian dynamics and the Aristotelian concept of force were the generally accepted notions in Muslim science. The only interesting exception, as far as the concept of force is concerned, is Abu Yusuf Ya'qub ibn Ishaq ibn al-Sabbah al-Kindi, "the philosopher of the Arabs." Alkindi combined Aristotelian concepts with Neo-Platonic notions and thereby contributed some new and original ideas to the concept of force. In his treatise on the tides[6] (*fi-l-madd wal-jazr*) his concept of force is still the traditional

[6] Eilhard Wiedemann, "Al-Kindi's Schrift über Ebbe und Flut," *Annalen der Physik* 67, 374–387 (1922).

Aristotelian *dynamis*. *Quwa* ("force" in Arabic) is for him both the potency to become warm, the susceptibility latent in the body (see p. 34), and also the "force" with which two bodies are struck to raise the fire.[7] However, most certainly under the influence of his optical investigations (his *De aspectibus,* a treatise on physiologic and geometric optics, was extremely influential on Roger Bacon, Witelo, and others), he conceived force as an entity propagated by rays. Not only light, and perhaps heat, are propagated by rays, according to Al-Kindi's view, but also every other type of force. He expounded his theory of rays in a treatise, *The theory of magic art* ("On stellar rays"), which in its Latin version appears in a number of medieval manuscripts, but unfortunately has never been described in full or newly edited.[8] His point of departure is the radiation of light from the stars, which carries also the astrological influences or forces, each star transmitting its own peculiar force through space. This idea of a parallelism between the transmission of light and the transfer of forces is actually in opposition to Aristotle's sharp differentiation between these two; for the propagation of light, as expounded by Aristotle in his *De anima*[9] and his *De sensu et sensato*[10] is instantaneous (an opinion accepted by all authorities in antiquity and the Middle Ages with the exception of Alhazen), whereas he explains in his *Physics*[11] that forces cannot act instantaneously, because a greater force would then be active in less than instantaneous time, which is an impossibility. Al-kindi, however, ignores this argument and conceives the propagation of forces in terms of the transmission of light. The motions of the stars and the "collisions of their rays" give rise to a great variety of combinations, so great, in fact, that no two objects of our sublunary

[7] *Ibid.;* see in particular p. 384 (misprinted in the *Annalen* as 484).

[8] See Lynn Thorndike, *A history of magic and experimental science* (Macmillan, London, 1923), vol. 1, p. 643.

[9] Aristotle, *De anima,* lib. II, cap. 7, 418a 20.

[10] Aristotle, *De sensu et sensato,* cap. 2, 438 a 12.

[11] Aristotle, *Physics,* lib. VI.

world can be alike in all their qualities. Moreover, not only the stars but all the elements of the world are sources of forces, fire, color and sound exemplifying the effusion of such forces.

Twelfth-century Christian scholarship was too much preoccupied with the issues of the problem of realism and nominalism to contribute much that is essential to the conception of force. Even astrological speculations on astral influences were not brought into the open until the school of Chartres, under the influence of their emissaries to Muslim Spain, found an interest in astrological thought. William of Conches' *De philosophia mundi,* although strongly inspired by Firmicus Maternus, shows many traces of Arabian astrological influences.[12]

Astrological speculations, and in particular Al-Kindi's conceptions, greatly influenced Roger Bacon's treatment of forces. For Bacon, every finite being is composed of matter and form and is created in time. The objects of the universe, though distinct from one another as to their substances, are united in an immense network of reciprocal actions and reactions, activated by the infusion of celestial virtues on the inferior bodies. To Bacon, believing in a finite spherical universe with the earth as its center, astrology was more than a working hypothesis for the explanation of the changes in the sublunary world. While celestial virtues were admitted as explanations by almost all Latin authors of the thirteenth century, Bacon was one of the first to inquire into their assumed physical mechanism. The fourth part of his *Opus maius,* and in particular his treatise *De multiplicatione specierum,* are dedicated to the clarification of the propagation of these forces. The central notion that Bacon here employs is "species," a concept much discussed in the literature on Bacon. It is a very obscure notion and Bacon himself apparently was unable to give an exact definition of it; nor were his successors, although many of them used it. Because of its paramount importance for Bacon's

[12] For a list of astrological writings of this period, see J. A. Clerval, *Les écoles de Chartres au moyen-âge* (Picard, Paris, 1895), p. 239.

conception of forces, let us quote the passage of his *Opus maius* in which he comes nearest to a definition of this concept:

In the first place, I shall demonstrate a proposition in geometry in respect to the efficient cause. For every efficient cause acts by its own force which it produces on the matter subject to it, as the light of the sun produces its own force in the air, and this force is light diffused through the whole world from the solar light. This force is called likeness, image, species, and by many other names, and it is produced by substances as well as accident and by spiritual substance as well as corporeal. Substance is more productive of it than accident, and spiritual substance than corporeal. This species causes every action in this world.[13]

The original Latin text follows:

Omne enim efficiens agit per suam virtutem quam facit in materiam subiectam, ut lux solis facit suam virtutem in aere, quae est lumen diffusum per totum mundum a luce solari. Et haec virtus vocatur similitudo, et imago, et species et multis nominibus, et hanc facit tam substantia quam accidens, et tam spiritualis quam corporalis. Et substantia plus quam accidens, et spiritualis plus quam corporalis. Et haec species facit omnem operationem huius mundi.[14]

In formulating thus a theory of causal action by adopting certain elements of the Neo-Platonic theory of emanations, Bacon calls the *influentia* by which the *agentia* act and react *similitudo,* since the virtue evoked in the *patiens* is similar to the virtue evoking it.

For further information on this concept let us turn to his treatise *De multiplicatione specierum.* Here Bacon states that species is not actually emitted from the agent. For if it were, the agent itself would be weakened thereby and would eventually waste itself and be consumed through the emission. The production of the species, on the other hand, is no *creatio ex nihilo,* but the agent excites the potential activity of the medium between itself and the patient and finally the potential activity of the substance of the patient. "Fit species de potentia activa materiae

[13] Roger Bacon, *The Opus maius,* trans. R. B. Burk (University of Pennsylvania Press, Philadelphia, 1928), vol. 1, p. 130.

[14] Bacon, *Opus maius,* pars IV, dist. II, cap. I.

patientis." The agent acts on that part of the medium or patient which is adjacent to it, and this part, in turn, transmits the species by stimulating the latent energy of the adjoining parts and so on. Every section activated by the species acts on the succeeding section, action producing reaction. "Omne agens physice patitur et transmutatur insimul dum agit, et omne patiens physice agit." [15] The transmission of force, in other words, is some kind of chain reaction that successively energizes consecutive parts of the medium or the patient. Species, however, is conceived by Bacon as something corporeal, its corporeal nature being identical with the corporeal nature of the medium. It seems to the present author that the nearest analogy in modern science is perhaps that of a wave which, like Bacon's species, is a form of transmission of energy or force, occupies time for its propagation, as does species, and has the corporeality of the medium involved.

For Bacon, however, more than this is involved. Species, in his view, is force detached from its subject, force outside the body in which it originated.

> In those beings that have reason and intellect, although they do many things with deliberation and freedom of will, yet this action, namely, a production of species, is natural to them, just as it is in other things. Hence the substance of the soul multiplies its own force, in the body and outside of the body, and any body outside of itself produces its own force, and the angels move the world by means of forces of this kind. But God produces forces out of nothing, which he multiplies in things; created agents do not do so, but in another way about which we need not concern ourselves at the present. Forces of this kind belonging to agents produce every action in this world.[16]

This passage, in which Bacon writes of forces as being outside the body, sounds as if he had a faint inkling of the law of the conservation of energy. In any case, he certainly subjects forces to a quantitative treatment in the second chapter, "in which the canons of the multiplication of the forces of agents as respects lines and angles are explained."

[15] Bacon, *De multiplicatione specierum* (Combachii, Frankfort, 1614).
[16] Bacon, *The Opus maius,* p. 130.

Every multiplication is either with respect to lines, or angles, or figures. While the species travels in a medium of one rarity, as in what is wholly sky, and wholly fire, and wholly air, or wholly water, it is propagated in straight paths, because Aristotle says in the fifth book of the *Metaphysics* that nature works in the shortest way possible, and the straight line is the shortest of all. This fact is also made evident by the twentieth proposition of the first book of Euclid, which states that in every triangle two sides are longer than the third.[17]

In the following sections Bacon speaks of the laws governing the propagation of forces through different media, expressing ideas that show some similarity to the notion of refraction of lines of force in modern electrostatics, for instance.

"Force," as we have seen, although carried and transmitted by the corporeal medium, is conceived by Bacon as an isolated entity, detached from the agent in which it originated. "Force" does not act at a distance but spreads through the medium according to specified laws of propagation. In this respect Bacon apparently follows certain ideas that were advanced in Greek science in order to explain vision and other optical phenomena. Aristotle in his treatise *On divination*,[18] refers to a theory of Democritus according to which a visible object continually emits so-called *eidola* that are able to affect the sleeper at night and account for his dreams. Lucretius, in his *De rerum natura* says: "Pictures of things and thin shapes are emitted from things off their surface."[19] Diogenes Laertius reports Epicurus as saying: "There are molds conforming to all solid bodies and preserving the same shape and configuration which emanate from them and spread through space with extraordinary speed. They may be called images."[20] Optical speculations of this type undoubtedly led Bacon to his mathematical treatment of forces and his conception of species.[21] No wonder that a major part of his *Opus maius*

[17] *Ibid.*, p. 131.

[18] Aristotle, *De divinatione per somnum*, cap. 2, 464 a 3.

[19] Lucretius Caro, *De rerum natura*, IV, 43–45, 155–160, 209–213.

[20] Cf. Diogenes Laertius, *Lives of eminent philosophers*, X, 46 (Loeb Classical Library; Harvard University Press, Cambridge, 1925), p. 575.

[21] It is with reference to such optical analogies that medieval epistemology used the term *species sensibilis* and later *species impressa*, through the reception

is devoted to optics, from which he hoped, in accordance with the metaphysics of light prevailing at his time, to obtain further information about the physical behavior of forces of all kinds. And indeed Bacon draws important conclusions from his theory of species as to the magnetic attraction, gravity, and attractive forces in general. The attraction of iron toward the magnet is frequently explained in the literature of the Middle Ages by Bacon's conception of species: the magnet evokes in its environment a *species magnetica* that spreads spherically through the medium, multiplies itself from one portion of the iron to the adjacent one until the iron receives in all of its parts the magnetic quality of tending to become united with the magnet, and *per accidens* gets subjected to local motion. The action at a distance is reduced thereby to a chain of contiguous contact-processes. By a similar modus operandi the tendency of a gravis to its natural place is explained.[22]

Bacon's theory of species had a great influence on Peter John Olivi,[23] the head of the Franciscan spiritualists at the end of the thirteenth century. In Olivi's discussion on causality the problem of action at a distance is raised and an attempt is made to solve it by means of the conception of species. In analogy to the process of vision in which *species visibiles* transmit the impressions from the agent to the patient, all causal processes are explained. "Primus effectus agentis est . . . eius similitudo," he says, just like Bacon. Olivi reduces action at a distance to action by contact by introducing the concept of species and submitting himself to the Aristotelian-Scholastic tradition, according to which "nihil agit in distans nisi prius agit in medium"; he seems,

of which the object was known. The formation of the concept of *species intelligibilis* (*species expressa*), on the other hand, led to intellectual abstraction, a process that culminated in the epistemology of Carolus Bovillus' *Liber de intellectu* (Paris, 1510).

[22] Roger Bacon, *Quaestiones supra libros quatuor physicos. Communia naturalium,* lib. I, pars III, dist. 2, cap. 3.

[23] Fr. Petrus Joannis Olivi, O.F.M., *Quaestiones in secundum librum Sententiarum, quas primum ad fidem Codd. Mss.,* ed. Bernardus Jansen, S.J. (Ad Claras Aquas Quaracchi, 3 vols., 1922–1926).

however, already to realize that contiguous transmission of causal action has its problems, too.

The problem of the motion of heavy bodies engaged many writers of the thirteenth century. An interesting theory was advanced by Bonaventure, who expressed numerous ideas that contradicted traditional Peripatetic conceptions. Thus, in his *Book of sentences*[24] he includes among the causes of the motion of a heavy body a force of repulsion exerted on it by the celestial spheres: "Dicendum quod ad motum gravis non sufficit solummodo gravitas sive qualitas propria, immo concurrit virtus loci attrahentis et virtus loci expellentis et virtus corporis quinti, praeter illa dua moventia, quae ponit Philosophus, scil. generans grave et leve, et removens prohibens." The last words refer to the Aristotelian explanation of gravity or levity, according to which the efficient cause is the production of the grave, that is, the impregnation of a portion of matter with the *forma terrae,* and the accidental cause is the removal of the obstacle or suspension that formed the resistance to the motion prior to its initiation. Bonaventure's hypothesis seems to have found no advocates, most probably because it assumed that a body would have to move with its greatest velocity (or, more accurately, acceleration) at the very beginning of its motion, it being tacitly assumed that repulsion attenuates with distance.[25]

The idea of attraction is met with also in the writings of Richard of Middleton, who said: "Mihi videtur dicendum, quod quamvis elementa ad suos motus naturales determinata sint a generante, tamen per virtutem suam et aliquam partipationem influentiae, quae viget in locis suis, exequuntur illos motus, ad quos a generante sunt determinata."[26] But even the assumption

[24] Doctoris Seraphici S. Bonaventurae, *In sententias,* II, dist. 14, pars I, art. III, qu. 2.

[25] Much later, however, in the early seventeenth century, the hypothesis that the attraction of terrestrial bodies decreases with decreasing distance from the center of the earth was regarded by some authors as a serious possibility. See, for example, F. Marin Mersenne's *Harmonie universelle contenant la théorie et la pratique de la musique etc.* (Cramoisy, Paris, 1636), p. 94.

[26] Ricardus de Mediavilla, *In sententias,* II, dist. 14, art. II, qu. 4.

of only a force of attraction, let alone the idea of repulsion, was not popular in those days. Two serious objections stood in the way of its general acceptance, two objections that medieval thought was unable to overcome. The first is that the assumption of an attractive force would lead to the conclusion that a body must be heavier in the immediately upper regions than farther above the surface of the earth, a conclusion that would stand in contradiction to the accepted traditional belief that a body is heaviest in the region of fire, less heavy in the region of air, and still less in water (a belief that is of course easily confirmable for air and water). The second objection is that if attraction was the *vis motrix* of a body, then a massy, weighty, bulky body should move with less speed than a small one, assuming that the resistance to motion is somehow proportional to the quantity moved. Thus, Albert of Saxony says in his commentary to *De caelo:* "Sequitur quod graviora tardius moverentur deorsum quam minus gravia . . . nam graviora plus resisterent virtuti trahenti." [27] It is here, perhaps for the first time, that a possible distinction, on an intuitive basis, between weight and mass is suggested. It was Newton who first recognized that not only the resistance, but also the magnitude, of the attractive force is proportional to the mass involved.

It was William of Occam who in his *Sententiae*[28] threw the concept of species overboard and acknowledged explicitly the concept of action at a distance. Rejecting Aristotle's principle of the immediate contact between the mover and the moved, he claimed that a *simultas virtualis* is quite sufficient to maintain the continuation of motion of the mobile detached from the mover, as the sun causes the "inferior light" ("quia sol causat lumen hic inferius") or as is illustrated by the attraction of the magnet. William's notion of an action at a distance did not find much acceptance among the schoolmen. Scholastic thought, on the whole, could not rid itself sufficiently of Peripatetic ideas

[27] Albert of Saxony, *De caelo,* II, qu. 14, art. 2.

[28] William of Occam, *Sententiae* (ed. Lyons, 1495), II, qu. 18.

to accept this revolutionary conception of force. Nevertheless, even in the traditional teachings of the schoolmen a major change in the conception of force took place: the gradual elaboration of the theory of gravity led eventually to the dismissal of the Aristotelian doctrine that every motion of inanimate objects requires a mover (*movens*) distinct and different from the mobile (*motum*). In conformity with Aristotle, the active principle that turns a body into a mover, that makes it capable of moving another object, is referred to as *virtus movens,* or *vis motrix.* Motive force is the cause of motion. If this force is constant, the mobile moves with constant velocity; if the force varies in its magnitude, the velocity is changed correspondingly. What forces were recognized by the Scholastics? An agent moves either naturally (*a natura,* corresponding to the Aristotelian *kata physin*) or it moves *ab intellectu* (*para physin*) mostly when moved by animal muscular forces. Consequently, apart from muscular forces, only natural forces were recognized, forces whose origin is found in the elementary qualities of the four elements: heat, cold, moisture, and dryness. In addition to these, gravity and lightness were accepted as motive forces; the primary qualities of the four elements were not supposed to initiate motion directly, but only through the intermediary of rarefaction or dilatation (*raritas*) and condensation (*densitas*). Yet, in this broader sense all six active qualities were conceived as efficient causes for local movement (*motus localis*). Whereas in Aristotle the nature of resistance to local motion is somewhat obscure, the Scholastics conceived resistance explicitly as a force (*vis resistiva*). As in Aristotle, resistance is a necessary element in the process of motion, for it is the cause of the temporal succession of motion (vis resistiva est causa successionis in motu). The velocity of the mobile is a function of the relation between the motive force and the resistance.[29]

[29] See my paper, "Physical thought of the fourteenth century and its contribution to the foundation of modern mechanics," *Koroth, A Quarterly Devoted to the History of Medicine and Science 1,* 5 (1953).

The exact form of this function, that is, the mathematical formulation of the law of motion, was a much disputed subject at that time. As explained on page 39, according to the traditional Peripatetic law of motion the velocity of the mobile is determined by the ratio of the motive force to the resistance: v is proportional to A/B. A different law of motion was advanced, toward the end of the sixth century A.D., by Joannes Philoponus,[30] who contended that v is proportional to the algebraic difference of the motive force and the resistance, that is, v is proportional to $A - B$. As Averroes relates in his commentary to "Text 71," Avempace (Abou Bekr Mohammed ben Ya'hya, commonly known as Ibn-al-Cayeg or Ibn-Badga) also asserted that v is proportional to $A - B$, basing his contention ultimately on Neo-Platonic metaphysical considerations.[31] The Neo-Platonic version of the law of motion (Philoponus, Avempace) was defended in the Middle Ages by Thomas Aquinas, Roger Bacon, Peter John Olivi, and Duns Scotus, whereas the traditional Peripatetic formula (Averroes) was supported by Albert the Great and Aegidius of Rome.

A third form of the law of motion was advanced by Thomas Bradwardine. In his *Treatise on proportions*[32] Bradwardine recognized that the traditional formulation has to be modified in order to include the case of an equality between the motive force A and the resistance B. Aristotle's original expression, if extrapolated for such conditions, leads to a nonzero velocity, a result inconsistent with experience. Bradwardine thus had to find a function of A/B that vanishes if this argument becomes unity. By a process of logical elimination, he arrives at his law of motion: "The proportion of the proportions of motive to resistive powers is equal to the proportion of their respective speeds of motion, and conversely. This is to be understood in the sense of

[30] Cf. Pierre Duhem, *Le système du monde* (Hermann, Paris, 1913), vol. 1, pp. 350–371.

[31] See pp. 50 ff.

[32] Thomas Bradwardine, *Tractatus de proportionibus* (1328), ed. and trans. H. L. Crosby, Jr. (University of Wisconsin Press, Madison, 1955), p. 113.

geometric proportionality."[33] In modern mathematical symbolism, Bradwardine's law of motion would read:

$$\text{velocity} = \log (A/B).$$

Bradwardine's introduction of this exponential function signifies an important step toward self-consistency in mechanics through mathematics, a movement toward a description whose deduced consequences do not contradict observational data in fields of experience beyond those originally referred to for the establishment of the formulation. H. L. Crosby, in the introduction to his translation of Bradwardine's *Treatise,* attempts to show that John Dumbleton's conception[34] of Bradwardine's law of motion is almost Newtonian. It seems to me, however, that Crosby's interpretation of the laws of "uniform intension of velocity" goes a little too far.

In accordance with the twofold nature of force, resistive forces, too, belong to two classes: external resistance (such opposing forces as pressure) exerted on the mobile in the opposite direction to its motion, and internal tendencies, particularly those that tend to return the mobile to its natural place. Gravity and lightness, therefore, belonging to the latter category, are conceived as forces. It is here that the theories advanced for an explanation of free fall begin to have a bearing on the conception of force. The problem of free fall, "haec quaestio inter omnes physicas quaestiones gravissima,"[35] was approached by the consideration that either the body moves by itself — a *movens motum* can, according to Aristotle, be only a living being — or it is moved by something else. Aristotle's exposition of free fall was no longer acceptable as a satisfactory explanation, since it ex-

[33] "Proportiones potentiarum moventium ad potentias resistivas, et velocitates in motibus, eodem ordine proportionales existunt, et similiter econtrario. Et hoc de geometrica proportionalitate intelligas." Bradwardine, p. 112. Cf. Marshall Clagett, *Giovanni Marliani and late medieval physics* (Columbia University Press, New York, 1941), pp. 125–144.

[34] John Dumbleton, *Summa logicae et philosophicae naturalis,* Cod. Vat. Lat. 954 and mss. Merton Coll. 306.

[35] Nicoletto Vernias, *De gravibus et levibus* (Venice, 1504).

plained only the initiation of the motion, but not its continuation.[36] Various new theories were advanced. In a majority of them a new conception of force was employed, the motion of heavy bodies being explained by an intrinsic principle, by a force supposed to be seated in the mobile itself. In analogy to the conception of impetus, a motive force conceived as inherent in the *projectum,* gravity was reduced to an intrinsic force in the *gravis,* a force that needs no further substratum for its action. Even when a more Platonic explanation of free fall was formulated, when it was assumed that like tends toward like, as in Nicolas Oresme's explanation,[37] the tendency as such was spoken of as an inherent inclination. The concept of force as an emanation of matter, in the Aristotelian sense, is replaced by the concept of force as an inherent activity, causing the motion of the body. The concept of force as inherent in the object is also employed by Duns Scotus,[38] John of Jandun,[39] and others. This concept conforms also with the traditional doctrine, expressed by Averroes and reformulated by Albert the Great, namely, that the weight of a *gravis* does not vary with its distance from the center of the earth, as long as it is within the boundaries of one and the same sphere of the elements (a doctrine rejected by Thomas Aquinas). For if the *vis motiva* is seated in the mobile itself, it is independent of its distance from the center of the earth, whereas an attraction at a distance would induce a

[36] Cf. Anneliese Maier, *Die Vorläufer Galileis im 14. Jahrhundert* (Storia e Letteratura, Rome, 1949), and her other publications.

[37] "Tunc dico quod motus localis gravis deorsum causae sunt 1. naturalis ordo et situs rerum, 2. forma sua seu natura sua, 3. gravitas quidquid dit illa, 4. contrarietas termini sive situs loci sibi naturaliter debiti, 5. forte removens obstaculum vel illud, quod ipsum ibi vel extra posuit, quia dat sibi quoddam impedimentum, 6. forte similitudo illius ad quod movetur, quia omnia similia habent quandam inclinationem naturalem adinvicem, sicut contraria habent se fugare et corrumpere naturaliter, nec de hoc potest reddi bene causa." Nicolas Oresme, *Quodlibeta,* qu. 22.

[38] Duns Scotus, *Sententiae,* II, dist. 2, qu. 10; *Metaphysics,* IX, qu. 14.

[39] John of Jandun, *Super octo libros Aristotelis de physico auditu* (ed. Venice, 1551), VIII, qu. II.

dependence on this distance. Copernicus still speaks of an intrinsic "appetition" of the mobile when he says:

> I think that gravity is nothing else than a certain natural appetition given to the parts of the earth by divine providence of the Architect of the universe in order that they may be restored to their unity and to their integrity by reuniting in the shape of a sphere. It is credible that the same affection is in the sun, moon, and other errant bodies in order that, through the agency of this affection, they may persist in the rotundity with which they appear to us.[40]

Parallel to the development of the concept of force as an inherent activity we have in the fourteenth century a different trend of ideas which led to an early anticipation of the notion of a *field* of forces. Curiously, it was John Buridan, one of the foremost proponents of the impetus theory, who formulated these ideas clearly:

> Debemus imaginari a toto caelo unam influentiam continuam usque ad centrum; tamen illa influentia prope caelum et remote habet aliam proprietatem et virtutem, et propter illam influentiam sic virtualiter diversificatam superius et inferius ordinant se gravia et levia in hoc mundo inferiori. Et non debet hoc negari ex eo quod illam influentiam non percipimus sensibiliter, quia etiam non percipimus illam quae de magnete multiplicatur per medium usque ad ferrum, quae tamen est magnae virtutis.[41]

It was assumed that the *virtus caelestis* permeates all space and thus exerts its influence on the bodies, more or less iike a stationary field of forces.

This theory of Buridan's is certainly not reconcilable with Aristotelian conceptions of causality: "Causa agens est simul cum suo effectu proximo et immediato." Still less consonant with Peripatetic thought is Buridan's rejection of intelligences (*intel-*

[40] "Equidem existimo gravitatem non aliud esse quam appetentiam quandam naturalem partibus inditam a divina providentia Opificis universorum, ut in unitatem integritatemque suam sese conferant in forman globi coeuntes." Nicolas Copernicus, *De revolutionibus orbium caelestium*, book 1, cap. 9.

[41] John Buridan, *Quaestiones super libris quattuor de caelo et mundo*, liber IV, qu. 2, ed. E. A. Moody (Mediaeval Academy of America, Cambridge, Publication No. 40, Studies and Documents, No. 6, 1942), p. 250.

ligences motrices) as the cause of the motion of the celestial spheres. In contrast to Aristotle and his medieval exponents who attributed the revolutions of the celestial bodies to such intelligences, Buridan advanced the theory that God, when creating the universe, communicated to these spheres an initial impetus, similar to that given to a stone when thrown. Although the impetus theory, originating perhaps with Philoponus and supported by several Muslim writers, has a long history even before Buridan, he was the first to apply this theory systematically to the circular motion of the stars:

> Nowhere does one read in the Bible that there exist intelligences charged with communicating to the celestial spheres their proper movements; it is therefore permitted to show that there is no need to suppose the existence of such intelligences. One can say that God, when creating the world, has moved, as he pleased, each of the celestial orbits; he has given to each of them an impetus which kept them moving since then . . . Thus he could rest on the seventh day from the work he has done.[42]

Buridan resumes this problem with the same solution at two other passages in his writings.[43]

Buridan's teachings were highly influential, especially in the Parisian nominalistic school. Thus Albert of Saxony, who left Paris in 1362, about four years after Buridan's death, accepted his theory of circular impetus:

> One can, indeed, maintain the following: When God created the celestial spheres, He put each of them in motion as He pleased; and they continue in their motion still today by virtue of the impetus which He impressed on them; this impetus is not subjected to any diminution,

[42] John Buridan, "Quaestiones octavi libri physicorum," Bibl. nat. fonds lat., ms. 14723, fol. 95, col. b, qu. 12, "Utrum projectum post exitum a manu projicientis poveatur ab aere, vel a quo moveatur."

[43] John Buridan, "Quaestiones quarti libri phisicorum," Bibl. nat. fonds lat., ms. 14723, fol. 68, col. c, "Quaeritur nono utrum in motibus gravium et levium ad sua loca naturalia tota successio proveniat ex resistentia medii"; "In Metaphysicen Aristotelis quaestiones argutissimae," *ibid.*, fol. 73, col. a, lib. XII, qu. 9, "Utrum quot sint motus coelestes, tot sint intelligentiae et econverso."

since the mobile has no inclination which could oppose the impetus, as no corruption there exists.[44]

Buridan's concept of a rotational impetus as the inherent cause of celestial movements was interpreted with the renewal of Platonic ideas in the fifteenth century in a more spiritualistic way. An excellent illustration of this reinterpretation is given in the writings of Nicholas of Cusa, cardinal and bishop of Brixen. In his *Dialogue on the sphere,* the duke of Bavaria, after a long discussion on the nature of spherical motion, asks the question: "But how did God create the motion of the outermost sphere?" The German cardinal answers:

> Much as you give motion to the globe. But this sphere is not moved directly by God, the Creator, nor by the Spirit of God; as it is not you nor your spirit who move immediately the globe which is now rotating in front of you. It is, however, you who initiate this motion, since the impulsion of your hand, following your will, produced an impetus and as long as this impetus endures the globe continues to move.[45]

In the continuation of this discussion the circular motion of the globe is compared to the soul inherent in the body. The impetus animates the globe as the soul animates the human body. In Cusa's *Dialogus trilocutorius de possest,* this comparison, or even identification, is further elaborated:

> The child takes the top which is dead, that is, is without motion, and wants to make it alive; . . . the child makes it move with rotational motion as the heavens move. The spirit of motion, evoked by the child, exists invisibly in the top; it stays in the top for a longer or shorter time according to the strength of the impression by which this virtue has been communicated; as soon as the spirit ceases to enliven the top, the top falls.

[44] Albert of Saxony, *Sublissimae quaestiones in libros de caelo et mundo,* lib. II, qu. 14. Cf. also his *Quaestiones in libros de physica auscultatione, libri octavi,* qu. 13.

[45] Nicolai de Cusa, *Dialogorum de ludo globi,* lib. I, trans. Gerda von Bredow; *Vom Globusspiel* (Schriften des Nikolaus von Cues, Meiner, Hamburg, 1952; Philosophische Bibliothek, vol. 233), and by Maurice de Gandillac, *Oeuvres choisies de Nicolas de Cues* (Aubier, Paris, 1942), p. 527.

In the sixteenth century the generally accepted concept of force — apart from "force" as employed in statics, considered as a push or a pull and never subjected to further analysis — was a rather eclectic combination of the principle of "like attracting like," the Neo-Platonic theory of cosmic sympathy, and the Peripatetic assumption of higher intelligences as the motive powers behind the stars.[46] Giordano Bruno[47] declares that there is no philosopher of importance who does not hold that the spheres of the universe are endowed with life and soul. Yet in the second half of this century stress was laid in particular on the theory of sympathy or cosmic harmony: Copernicus' interpretation of gravity, in 1543, as a tendency or "appetition" of the parts to be united with the whole to which they belong, was in the last analysis a theory of sympathy. This conception of force as an inherent tendency or propensity to be united with the cognate is congenial with the notion of force as the emanation of a world soul. Both conceptions stem from the ancient Platonic tradition, revived in the school of Chartres, and transmitted through Cusanus, Ficinus, and other thinkers right down to Copernicus.[48]

Now, it would lead us too far into details should we give even an abbreviated list of sixteenth-century publications that propounded a theory of force along these Platonic lines. As examples of the trend of the ideas of this period we shall mention only two works and study their influence upon the subsequent intellectual development.

A few years after the publication of Copernicus' *De revolutionibus orbium coelestium,* a most detailed theory of sympathy of parts to the whole was elaborated by Girolamo Fracastoro. The

[46] For a typical document of this conception, see Cesare Cremonini, *Disputatio de coelo in tres partes divisa* (De natura coeli. De motu coeli. De motoribus coeli abstractis . . .) (Venice, 1613).

[47] Giordano Bruno, *De la causa, principio et uno* (Charlewood, London, 1584).

[48] For details about the continuity of this tradition, see Raymond Klibansky, *The continuity of the Platonic tradition during the Middle Ages* (Warburg Institute, London, 1950).

following quotation from this work shows how this author conceived the mechanism of sympathetic activity.

When two parts of the same whole are separated from each other, each sends toward the other an emanation of its substantial form, a species propagated into the intervening space; by the contact of this species each of the parts tends toward the other in order to be united in one single whole; this is the way to explain the mutual attraction of like to like, the sympathy of iron for the magnet being a typical example.[49]

The same mechanism comes into operation also if two parts or substances are related to each other by a common quality, as are fire and light by dryness, and the Aristotelian conception of a body's tendency to its "natural place" is interpreted by Fracastoro as such operation of sympathy. Indeed, all motion is determined by the sympathetic or antipathetic relations of the moving body with respect to its environment. The third chapter of Fracastoro's *De sympathia et antipathia rerum,* carrying the chapter title "De sympathijs elementorum ad loca propria," discusses this problem in detail. For example, in it Fracastoro argues that the "natural place" of fire is between the concave lunar sphere on the one hand and the convex upper boundary of the atmosphere on the other, because fire is sympathetically related to light, which occupies the place above the lunar sphere. The tendencies of all four elements are similarly interpreted as resulting from their sympathetic relations to their respectively adjacent spheres and elements. "Terra vero circa centrum in aquae concavo locum habet, cui frigiditate respondet." [50]

A second interesting document of this late Neo-Platonic conception of force is Antonius Ludovicus' *On occult properties,*[51] published in Lisbon in 1540. Ludovicus, or, as he is often called,

[49] Girolamo Fracastoro, *De sympathia et antipathia rerum* (Venice, 1546), lib. I; reprinted in *Hieronymi Fracastorii Opera omnia* (Venice, 1555).

[50] *Hieronymi Fracastorii Veronensis Opera omnia* (secunda editio, Apud Iuntas, Venetiis, 1573), p. 59.

[51] *Antonii Lodovici medici Olyssipponensis De occultis proprietatibus libri quinque, Opus praeclarissimum* (Lisbon, 1540).

Luiz,[52] a "great physician and philosopher"[53] and professor at the university of Coimbra, was hailed by his compatriots[54] in the eighteenth century as a precursor of Newton. Even Lemos,[55] who gives a detailed critical analysis of Ludovicus' works, comes to the conclusion: "We are not justified in charging the great British discoverer with plagiarism as Soyé does but we should regard our fellow-countryman as his precursor."[56] But before drawing any hasty conclusions, let us see what Ludovicus has to say for himself. In the preface to the second book of his work he says:

> This faculty of attraction is in fact apparent far and wide, in seeds, in plants, in metals as well as in animals. And finally I would dare to assert that throughout the whole of nature a certain faculty of attraction is diffused which ties together its separate parts in an indissoluble connection. For it is not easy to find some object which does not exhibit a friendly relation to something else or is not separated by the mutual communion of nature, which sympathy [*convenientia;* see p. 44] or antipathy [*disconvenientia*] produces attraction, as we shall demonstrate. Owing to this virtue the universe itself is bound together, and so are its parts by invisible ties; however far they are removed from each other, they are bound together not to be dispersed.[57]

Thus, he continues, "similia similibus conjugantur," like joins like; this faculty of attraction maintains order in the universe.

[52] For his biography see *Enciclopedia universal ilustrada* (Espasa, Barcelona), vol. 31, p. 681.

[53] Barbosa Machado, *Bibliotheca Lusitana* (Lisbon, 1747), vol. 1, p. 311.

[54] See, for example, Luiz Raphael Soyé, *Sonho, poema erotico* (Lisbon, 1786), p. 14.

[55] Maximiano Lemos, *Archivos de historia da medicina Portugueza* (Porto, 1916), vol. 1, pp. 102–107.

[56] Maximiano Lemos, *Historia da medicina em Portugal* (Lisbon, 1899), vol. 1, p. 311.

[57] "Latissime autem haec attractix facultas patet in seminibus, in plantis, in metallis, in animalibus. Et denique ausim affirmare attracticem quandam facultatem, per omnem naturan diffusam esse, quae singula nexu indissolubili devinciat. Nec enim aliquam rem reperire quis facile possit, quae non ad aliam quampiam: vel amicam familiaritatem habeat, vel naturae communione non dissideat, ex qua convenientia, vel disconvenientia attractiones fieri docebimus. Per hanc virtutem mundus ipse connectitur, et mundi partes invisibilibus nodis: quamvis longissime distantes, ne diffluant continentur." Lodovici, *De occultis proprietatibus,* lib. 2, prohoemium, p. 16.

And before going into further details he thinks it necessary to show where these forces of attraction are most manifest. Consequently, the first chapter of the second book is devoted to a discussion of the magnet, the following chapters to the attractive forces exerted by grain, lupine, and cress.

These later chapters show clearly that Ludovicus, in his assumption of an attractive force for the explanation of physiological processes, closely follows Galen, who in his book *On natural faculties*[58] reduced such processes to the exertion of an attractive force (*dynamis helktike*), a concept suggested already in the Hippocratic writings. As to Ludovicus' theory of universal ties, binding the individual parts of the world into one coherent unity, Ludwig Edelstein remarks:

> The "certain power of attraction" which is conceived as a cosmic cause seems to be identical with what is called by Stoics and Neo-Platonists "sympathy," or by Sceptics "necessity of nature," through which all things are held together. Such a belief, again, was very common from the classical era down to Ludovicus' time. Whether the idea of attraction, as proposed by Ludovicus, is in any way an anticipation of Newton's law of attraction, can hardly be decided. No less a philosopher than Berkeley interpreted the Hippocratic-Galenic dynamism in this way. Yet the ancient concept lacks every mathematical connotation and is a metaphysical rather than a scientific term.[59]

The present author concurs with Edelstein's assessment and regards Ludovicus' work not as an original contribution of new ideas but rather as a publication in one line with the works of Fracastoro, Cardan, and many other sixteenth-century writers. However, these works, though rooted in Neo-Platonic conceptions, transform the more or less mythical conception of a world soul into a more scientific idea, namely, that the various parts of the universe, by being parts of one organic whole, influence each other mutually. It is this climate of opinion that led the

[58] Galen, *On the natural faculties,* trans. A. J. Brock (Loeb Classical Library; Harvard University Press, Cambridge, 1928).

[59] Ludwig Edelstein's remarks are printed in Harry Friedenwald, *The Jews and medicine* (Johns Hopkins University Press, Baltimore, 1944), vol. 1, p. 325.

sixteenth century to the conception of an immanence and self-sufficiency of the laws of nature, precluding thereby the possibility of direct intervening influences of a human or demonic kind. It led to the age of the so-called "natural magic," which tries to subdue nature, not by power of the word or the symbol, but by the command of nature's own regular inherent potentialities and capacities, an intellectual development which culminated in Giambattista della Porta's *Magia naturalis,* published in 1589, one of the early manifestoes proclaiming the scientific spirit of the Newtonian era. Only in this more reserved and more profound sense can Ludovicus, like many other thinkers of his time, be called a precursor of Newton.

The conception of nature as an autonomous, self-sufficient organism with its own and immanent laws was reflected in the assumption of a vast multitude of supposed forces that, owing to their qualitative and quantitative indeterminateness, had the character of merely occult qualities. Although the ancient and aesthetic conception of nature, which emphasized the opposition of form and substance, was gradually replaced by an opposition of force and matter, the conception of force still lacked all the quantitative and even physical determination necessary to form a sound foundation for a new physical science.

An important exception, however, was Bernardino Telesio and the Academy of Cosenza which he founded. Discarding the Aristotelian notion of form, he systematically introduced forces as the *principia agentia* and searched for a correlation of these forces (and this is the important point) with immediate experience. In his *De rerum natura juxta propria principia,*[60] published in 1565, he reduces all active forces to the force of expansion by heat and the force of contraction by cold. All changes and variations in nature are, in his view, but participations of the corporeal matter (*corporea moles*) in these two forces of heat and cold to different degrees.

[60] *Bernardini Telesii Consentini De rerum natura juxta propria principia liber primus et secundus* (Apud Antonium Bladum, Rome, 1565). The work was republished in Naples in 1570 and (enlarged to nine books) in 1586.

As explained in the first four parts of the work, heat and cold are the primordial forces, prior to the creation of heaven and earth. In fact, their contest for supremacy led to the creation of heaven and earth, which then became the sources of heat and of cold respectively. In the view of this most important representative of Italian natural philosophy in the Renaissance, heat is the originator of motion and life, cold of rest and rigidity. In the fifth part of his work he even identifies spirit (*spiritus*) with some kind of heat and applies his hypothesis to psychology and epistemology. What is important for the history of our subject is his insistence that heat is prior in both time and nature to motion.[61] This assertion, in contrast to many former statements[62] which claimed the precedence of motion to heat, if successfully worked out and divested of all unnecessary metaphysical connotations, would have led to a conceptual scheme of physics in which some thermal theory would have been the basic discipline, much as mechanics was for the physics of the nineteenth century. But it was, of course, the inaccessibility to mathematical treatment of Telesius' forces of heat and cold that led to their immediate rejection as basic conceptions in science.

Similarly, the great difference between the conception of a universal force as advanced, for example, by Ludovicus and Newton's universal attraction was, of course, the complete absence of any mathematical determination with Ludovicus. His concept of force was a purely metaphysical idea, whereas with Newton force became a quantity subjected to mathematical rules, thereby gaining methodological and technical importance.

There was, however, a much earlier attempt to geometrize the concept of force, although along a completely different line from Newton's approach: the points of application of the force were considered as mathematical entities, rather than the magnitude

[61] *Ibid.*, book I, chapter 8.

[62] See, for example, Max Jammer, *The history of science, an introduction to the study of the historical development of scientific thought* (Kiryat Sepher, Jerusalem, 1950), p. 109.

of the force. The tendency of a terrestrial object to approach the earth was conceived as acting between their centers of gravity rather than as existing between the parts of these two bodies. Albert of Saxony, referring to certain passages in Simplicius, Marsilius of Inghen, Peter of Ailly, and Nifo, alluded to such a possibility,[63] and explicit formulation of this conception is found in Guido Ubaldo del Monte's *Paraphrasis:* "When we say that a heavy body tends by a natural propensity to place itself in the center of the universe, we mean to say that this heavy body's own center of gravity tends to be united with the center of the universe."[64] These ideas, although aimed at a more quantitative and mathematical determination of the otherwise vague and undefinable forces of attraction, invested space, or, more accurately, geometric points in space, with an occult quality of attracting material particles or their centers of gravity. These conceptions, as we shall see, became the target of severe attacks by William Gilbert and Johannes Kepler, two important precursors of the new concept of force of the seventeenth century.

Even such an experimentally minded physicist as William Gilbert could not rid himself of the conception of a sympathetic relation of the parts to the whole. Though rejecting occult qualities in the explanation of physical phenomena, he still had to adhere to psychic elements in his exposition of attractive forces.

> But simple right-downward motion assumed by the Peripatetics is the movement of weight, of coacervation, of separated parts, in the ratio of their matter, by right lines toward the earth's center, these tending to the center by the shortest route.[65] We will now show the reason of its coition and the nature of its motion . . . It is not what the Peripatetics call causa formalis and causa specifica in mixtis and secunda forma; nor

[63] Cf. Pierre Duhem, *Les origines de la statique* (Hermann, Paris, 1905–1906), chap. 15, "Les propriétés mécaniques du centre de gravité."

[64] Guido Ubaldo del Monte, *In duos Archimedis aequiponderantium libros paraphrasis scholiis illustrata* (Pisa, 1588), p. 10.

[65] William Gilbert of Colchester, *On the loadstone and magnetic bodies and on the great magnet, the earth. A new physiology,* trans. P. F. Mottelay (Wiley, New York, 1893), p. 335.

is it causa propagatrix generantium corporum; but it is the form of the prime and principal globes; and it is of the homogeneous and not altered parts thereof, the proper entity and existence which we may call the primary, radical, and astral form, not Aristotle's prime form, but that unique form which keeps and orders its own globe. Such form is in each globe — the sun, the moon, the stars — one.[66]

In his *Philosophia nova,* which was published posthumously in 1651, Gilbert emphasizes the psychic origin of the attractive forces in stronger words:

Everything terrestrial is united to the earth; similarly, everything homogeneous with the sun tends toward the sun, all that is lunar toward the moon, and the same applies to the other bodies constituting the universe . . . It is not a question of an appetite which brings the parts toward a certain place, a certain space, a certain term, but of a propensity toward the body, toward a common source, toward the mother where they were begotten, toward their origin, in which all these parts will be united and preserved, and in which they will remain at rest, safe from every peril.[67]

The conception of a reunion, as formulated by Gilbert, or of an *appetentia,* as advanced by Copernicus,[68] is of course a logical conclusion suggested by the so-called Copernican revolution. If the earth is really just one of the planets revolving around the sun, and if the sun, as claimed by Bruno, is just one of the many stars, then gravity as experienced on the surface of the earth must be effective also on other planets, and, moreover, the

[66] *Ibid.,* p. 105. In the original we read: "Forma illa singularis est, et peculiaris, non Peripateticorum causa formalis, et specifica in mixtis, et secunda forma, non generantium corporum propagatrix; sed primorum et praecipuorum globorum forma; et partium eorum homogenearum, non corruptarum, propria entitas et existentia, quam nos primariam, et radicalem, et astream appelare possumus formam; non formam primam Aristotelis, sed singularem illam, quae globum suum proprium tuetur et disponit. Talis in singulis globis, Sole, luna, et astris, est una." (Guilielmi Gilberti Colcestrensis, *De magnete magneticisque corporibus et de magno magnete tellure, physiologia nova* (Petrus Short, London, 1600), p. 65.

[67] "Non autem est appetitus aut inclinatio ad locum, aut spatium, aut terminum; sed ad corpus, ad fontem, ad matrem, ad principium, ubi uniuntur, conservantur, et a periculis vagae partes revocatae quiescunt omnes." Guilielmi Gilberti Colcestrensis, *De mundo nostro sublunari philosophia nova* (opus posthumum; Elzevir, Amsterdam, 1651), p. 115.

[68] See reference 40.

Aristotelian differentiation between terrestrial motions in straight lines and celestial motions in circles has to be discarded. Gravity must be a property common to all celestial bodies, each body exerting its own attractive forces on its "cognate" parts. Prior to Kepler's revolutionary generalizations, which will be discussed in the next chapter, these attractive forces were conceived as characteristic of each celestial body separately and as confined to the proper sphere of activity of each individual star.

In conclusion of the present discussion it should be noted that the concepts of "attraction of the parts by the whole" and of "tendency toward reunion" are almost equivalent notions in the writings of all sixteenth-century authors. With Bullialdus, Galileo, and Borelli, however, the former concept implies the action of an external force, whereas the latter implies that of an internal one.

CHAPTER 5

THE SCIENTIFIC CONCEPTUALIZATION OF FORCE: KEPLER

A decisive stage in the development of our concept is reached with Johannes Kepler in his effort to attain a mathematical formulation and accurate determination of force. What matters is not the result achieved — from the viewpoint of Newtonian mechanics his conclusions were erroneous — but his new approach, his search for a *quantitative* definition of force.

In our analysis of this important stage we shall try to adhere as much as possible to the chronological order of Kepler's works and letters, although Kepler's intellectual development was a process of extreme vacillations. Indeed, in one and the same work the concept of force is sometimes referred to as a soul and sometimes as a physical, almost mechanical corporeal quantity. But, as Macaulay remarked in another context, vacillation cannot be regarded as a proof of dishonesty. On the contrary, Kepler's inconsistency of expression reflects only the extreme intellectual sincerity of a profound thinker who, unintentionally and to his own surprise, laid the foundations of a new approach and a new conceptual scheme, radically different from the scholastic edifice of thought.

In his *Mysterium cosmographicum*[1] Kepler still maintains the

[1] Johannes Kepler, *Prodromus dissertationum cosmographicarum continens mysterium cosmographicum de admirabile proportione orbium coelestium* . . . (Tübingen, 1596. Cf. *Guilielmi Gilberti Colcestrensis De Magnete magneticique corporibus, et de magno magnete tellure* (London, 1600), book V, chapter XII; "Vis magnetica animata est, aut animatam imitatur, quae humanam animam dum organico corpori alligatur, in multis superat."

traditional conception of force as a soul animating the celestial bodies and directing their proper motions. However, already in this early work a faint anticipation of some kind of force emanating from the central body seems to be in the back of his mind when he says: "The moon follows or rather is drawn (*trahitur*) wherever and however the earth moves along. Imagine the earth at rest, and the moon will never find her path around the sun." [2] It also seems that in certain passages in this work Kepler employs the term "soul" (*anima*) merely as a metaphor to express the immateriality of the principle that governs the mutual movements of the heavenly bodies. He simply is not yet in possession of a special term for this notion. Yet, in his first work to be printed in Prague, in 1601, *De fundamentis astrologiae certioribus,* he says again: "The same reason which induced the ancients to posit a third soul in the planets, compels us to posit a fourth soul in the earth." [3] However, his letters, and in particular his correspondence with David Fabricius, bear witness that he was not satisfied with such assumptions.

In a letter of March 28, 1605, addressed to Herwart of Hohenburg, Kepler conceives the universal nature of gravitational forces and calls gravity a "passivity" rather than an activity, approaching thereby a more functional conception than a psychic one. "If we were to place the earth at rest in some place and bring near to it a larger earth, the first would become a heavy body in relation to the second and would be attracted by the latter just as a stone is attracted by the earth. Gravity is not an action but a passivity of the stone which is attracted." [4] The

[2] "Luna sequitur vel trahitur potius, quocunque Tellus quacunque varietate graditur. Finge Tellurem quiescentem, nunquam Luna viam circa Solem inveniet." Kepler, *Mysterium cosmographicum,* cap. XVI, in *Gesammelte Werke,* ed. Max Caspar, vol. 1 (Beck, Munich, 1938), p. 55.

[3] "Et quae ratio veteres coegit tertiam animae specimen in plantis collocare, eadem nos cogit, quartam hanc collocare in terra." Kepler, *Gesammelte Werke,* ed. Max Caspar and Franz Hammer, vol. 4 (Beck, Munich, 1941), p. 23.

[4] "Si autem ad Tellurem quocunque in loco quiescentem applicaretur Tellus alia, et major, tunc illa sane fieret gravis respectu majoris, attraheretur enim ab illa, plane uti haec Tellus lapides attrahit etc. Itaque gravitas non est actio sed passio lapidis, qui trahitur, principium inquam ejus." Kepler, *Gesammelte*

reciprocity of attraction and mutual approach is clearly expressed in a letter of October 11, 1605, addressed to David Fabricius, in which Kepler states that not only does the stone approach the earth but also the earth approaches the stone.[5] That the forces of attraction belong to the material aspects of reality and consequently are subject to the mathematical formalism is stated by Kepler in a letter of November 30, 1605, addressed to Johann George Brengger.[6]

Of great importance for the understanding of the historical context in which Kepler arrived at his conception of force is another letter, also addressed to Herwart and written toward the end of January 1607. Here Kepler refers to Franciscus Patritius and his discussion on the tides. In the twenty-eighth book of his *Pancosmia*[7] Patritius gives a detailed history of the theories about the causes of the tides, naming in particular Federicus Chrysogonus who studied most accurately the time relation between the tides and the motion of the moon and came to the conclusion of a temporal coincidence of the tides with certain lunar positions. As a result of these investigations, Patritius relates, Chrysogonus, Fredericus Delfinus, Augustinus Caesareus, and finally Telesio advanced the theory that the moon causes the tides. Kepler agrees with these advocates of the moon theory, but generalizes these findings in an important sense: the oceans are attracted by the moon much as all heavy objects, the oceans included, are attracted by the earth.[8] Although

---

Werke, ed. Max Caspar, vol. 15 (Beck, Munich, 1951), Letters, 1604–1607, letter No. 340, p. 184.

[5] "Non tantum lapis ad Terram eat sed etiam Terra ad lapidem." *Ibid.,* p. 241.

[6] "Posses Sympathiam magneticam respectu Sympathiae coeli et Terrae dicere materialem et corporalem." Kepler, *Gesammelte Werke,* ed. Max Caspar, vol. 16 (Beck, Munich, 1954), Letters 1607–1611, p. 84.

[7] Franciscus Patritius, *Nova de universis philosophia libris quinquaginta comprehensa* (Mammarellus, Ferrara, 1591).

[8] "Aliquammulta de hac re in Franciso Patricio inveniuntur: quamvis perperam is Lunam (si bene meminj) excludere conatur a consideratione causarum. Ex illo Germano authore nata mihi est haec speculatio: a Luna maria sic attrahj, ut gravia omnia, ipsaque maria, attrahuntur a terra." Kepler, *Gesammelte Werke,* vol. 15, p. 387.

it is, as Kepler calls it himself, only a speculation and may be interpreted as if Kepler meant to say that lunar attraction works in analogy to terrestrial gravity, yet it seems, if taken in the context of his other works, that Kepler had already conceived the universal character of attraction, an idea generally attributed only to Newton.

To substantiate our interpretation by further evidence, it should be pointed out that Kepler, following his teacher Maestlin, was convinced of the physical similarity between the earth and the moon and before Galileo had directed his telescope to the surface of the moon, Kepler, in his *Mysterium cosmographicum,* accepted Maestlin's thesis that the physical conditions on the moon and on the earth are alike.[9] It was, therefore, not a far cry for Kepler to infer from a similarity of appearance to a similarity of behavior. If our interpretation is correct — and there seems to be every reason to believe that it is — it is Kepler and not Newton who has to be credited with having suggested "that the familiar behavior of falling bodies and the majestic sweep of the moon in its orbit were all part of the same great scheme." Newton, however, was the first to demonstrate clearly the correctness of this assertion.

In his letter to David Fabricius of November 10, 1608, Kepler clearly envisages the forces of attraction exerted by the earth on a stone as magnetic lines, or chains, as he says, thereby approaching Gilbert's conception of gravity as a magnetic emanation. Rejecting Tycho Brahe's argument, that the earth cannot rotate in space because bodies projected into the air return to their points of departure, Kepler attempts to explain this as follows:

> How is it possible that a sphere, thrown vertically upward — while the earth rotates meanwhile — does return to the same place? The answer is that not only the earth, but together with the earth, the magnetic

[9] "Atque ut spacium Luna ex orbe terreno, motumque sortita est, sic et multas conditiones globi terreni adeptam, puta, continentes, maria, montes, aerem, vel his aliqua quocumque modo correspondentia, multis conjecturis Maestlinus probat." *Mysterium cosmographicum,* cap. XVI, *ibid.,* vol. 4, pp. 55–56.

invisible chains rotate by which the stone is attached to the underlying and neighboring parts of the earth and by which it is retained to the earth by the shortest, that is, the vertical line.[10]

It is, however, more the motion of the planets with reference to the sun than the motion of a terrestrial object that attracts his attention. In fact, the search after a dynamical explanation of the planetary motion is one of the major problems already in his first great astronomical treatise, the *Astronomia nova*,[11] the "first modern book in astronomy."[12] In the introduction to his monumental work he investigates the nature of gravity:

Here is the true doctrine of gravity. Gravity is a mutual affection among related bodies which tends to unite and conjoin them (of which kind also the magnetic faculty is), while the earth attracts the stone rather than the stone tends toward the earth. Even if we placed the center of the earth at the center of the world, it would not be toward this center of the world as such that heavy bodies would be carried, but rather toward the center of the round body to which they are related, that is toward the center of the earth. Thus, no matter whereto the earth is transported, it is always toward it that heavy bodies are carried, thanks to the faculty animating it.[13]

One thing is certain: a mathematical point cannot exert that force of attraction. "Punctum mathematicum . . . nequit movere

[10] "Cur globus sursum missus ad perpendiculum, recidat ad locum eundem, si Terra interim abit. Respondendum enim, non tantum Terram interim abire, sed unam cum Terra etiam cathenas illas Magneticas infinitas et invisibiles, quibus lapis alligatus est ad partes Terrae subjectas et circumstantes undique, quibus retrahitur proxima id est perpendiculari via ad Terram. Quemadmodum igitur hic, vis infertur cathenis illis a motu violento sursum, quo fit ut omnes illae aequaliter quasi extendantur: Ita quoque vis infertur cathenis occidentalibus, cum globus violentia in orientem truditur, et vis infertur orientalibus, cum vapor globum proruit in occidentem." *Ibid.*, vol. 16, p. 196.

[11] *Joannis Kepleri Astronomia nova aitiologētos seu physica coelestis, tradita commentariis de motibus stellae Martis* (Heidelberg, 1609).

[12] Max Caspar, *Johannes Kepler* (Kohlhammer, Stuttgart, 1948), p. 159.

[13] "Gravitas est affectio corporea, mutua inter cognata corpora ad unitionem seu conjunctionem (quo rerum ordine est et facultas magnetica) ut multo magis Terra trahat lapidem, quam lapis petit Terram. Gravia (si maxime Terram in centro mundi collocemus) non feruntur ad centrum mundi, ut ad centrum mundi, sed ut ad centrum rotundi cognati corporis, Telluris scilicet. Itaque ubicunque collocetur seu quocunque transportetur Tellus facultate sua animali, semper ad illam feruntur gravia." *Joannis Kepleri Astronomia nova, Introductio,* in Kepler, *Gesammelte Werke,* vol. 3, p. 25.

gravia,"[14] he states emphatically. Since force presupposes for its activity the existence of an animated being, its source must be an extended, physical object animated with this faculty.

Kepler here agrees with Gilbert who contended in his *De magnete* that

it is in bodies themselves that acting force resides, not in spaces or intervals. But he who thinks that those bodies are at leisure and keeping holiday, while all the virtue of the universe appertains to the very orbits and sphaeres, is on this point not less mad than he who, in some one else's house, thinks that the walls and floors and roof rule the family rather than the wife and thoughtful paterfamilias.[15]

For Kepler at this time, as we see, force is still an animating faculty, which expression he uses for lack of a proper word to express the immateriality of its essence. In fact, the *anima motrix* of the *Mysterium cosmographicum* is now, in his *Astronomia nova,* a *species immateriata.*[16] This immaterial species, however, is already susceptible of mathematical treatment.

If two stones were removed to some place in the universe, in propinquity to each other, but outside the sphere of force of a third cognate body, the two stones, like magnetic bodies, would come together at some intermediate place, each approaching the other through a distance proportional to the mass (*moles*) of the other.[17]

In the third part of the *Astronomia nova* Kepler expresses in an unequivocal manner his conviction that these forces are subject to mathematical rules when he says:

[14] "Punctum mathematicum, sive centrum mundis sit sive non, nequit movere gravia neque effective neque objective, ut ad se addedant." *Ibid.,* p. 24.

[15] William Gilbert, *On the loadstone* [trans. S. P. Thompson] (Chiswick Press, 1900, London), book VI, chap. III, p. 217. The original Latin text reads: "In corporibus ipsis vis agens existat, non in spatijs, aut intervallis." Guilielmi Gilberti Colcestrensis, *De magnete* (Petrus Short, London, 1600), p. 217. In his *De mundo nostro sublunari philosophia nova* (opus posthumum) (L. Elzevir, Amsterdam, 1651), Gilbert says likewise: "Locus . . . vim non habet; potestas omnis in corporibus ipsis. Non enim Luna movetur nec Mercurii, aut Venus stella, propter locum aliquem in mundo."

[16] "Flumen est species immateriata virtutis in Sole magneticae." Kepler, *Astronomia nova,* vol. 3, p. 350.

[17] "Si duo lapides in aliquo loco mundi collocarentur propinqui invicem, extra orbem virtutis tertii cognati corporis, illi lapides ad similitudinem duorum magneticorum corporum coirent loco intermedio, quilibet accedens ad alterum tanto intervallo, quanta est alterius moles in comparatione." *Ibid.,* p. 25.

For we see that these motions take place in space and time and this virtue emanates and diffuses through the spaces of the universe, which are all mathematical conceptions. From this it follows that this virtue is subject also to other mathematical "necessities." [18]

With the progressing elaboration of his laws of planetary motion, and in particular with his recognition that the planetary velocity is greatest at perihelion and least at aphelion, his conviction becomes stronger from day to day that the immaterial faculty, seated in the sun and responsible for these motions, is something mechanical rather than spiritual. In fact, in his *Epitome astronomiae Copernicanae,* perhaps the first true treatise on celestial mechanics, Kepler already conceives this faculty as a force (*vis*) [19] in the mechanical sense of the word,[20] although this conception is naturally still intermingled with spiritual elements of expression. In other words, considerations of a methodological nature — the determination of the dependence of planetary velocity upon distance — were what led Kepler to the conclusion that the motive cause of these phenomena is something of an essentially physical nature.

In fact, from his study of Tycho Brahe's exceedingly accurate observations of planetary motion Kepler discovered that the planets did not move with uniform speed along their orbits, but traveled more slowly the greater their distance from the sun, and more rapidly the nearer they came to the sun. This relation is implied in what we usually call today "Kepler's second law," which states that the radius vector joining the sun and any given planet sweeps out equal areas in equal times. Kepler's own formulation of this law — which he discovered, incidentally, before the so-called "first law" — was slightly different, and strictly speaking was correct only at the apsides where the

[18] "Videmus enim motus istos perfici in loco et tempore, et emanare atque diffundi virtutem hanc a fonte per spacia mundi; quae sunt omnia res Geometricae. Quin igitur et caeteris Geometricis necessitatibus obnoxia sit haec virtus." *Ibid.*, p. 241.

[19] "Vis seu energia." *Joannis Kepleri Epitome astronomiae Copernicanae,* lib. IV, in Kepler, *Gesammelte Werke,* vol. 7, p. 547.

[20] "Sol . . . omnibus sui corporis partibus facultatem hanc activam et energeticam possidet attrahendi vel repellendi vel retinendi planetam." *Ibid.*

radius vector is perpendicular to the tangent; it stated that the velocity of a given planet is inversely proportional to its distance from the sun. Thus confronted with the discovery of a mathematical relation that holds throughout space, Kepler, in his search for deeper causes, was led to the assumption of the existence of a regulative force, attributing the fluctuations in speed to corresponding fluctuations in the magnitude of their physical correlate (force). Although Kepler's computations, it is true, were confined almost exclusively to the motion of the planet Mars, he rightly anticipated that his conclusions were valid for all planets without exception.

It is most interesting to note that at this critical point, where the concept of force is introduced into scientific methodology, Kepler discusses a problem of scientific method. He says that if two phenomena occur simultaneously and in the same manner, and if they exhibit always the same relation, it is an axiom of scientific method to regard one of them as the cause of the other or to regard both as resulting from the same cause. Therefore, the decrease of velocity is the cause of the increase of distance or vice versa; it is also possible that both have a common cause. Kepler holds that the increase of distance is the cause of the decrease of velocity.[21] But at this stage, Kepler is not yet satisfied in his desire for causal explanation. How can mere change of distance from the center have an effect upon the velocity of the moving planet? Kepler thus introduces an intermediary concept (a construct) as a bridge, so to say, between change of distance and variation of speed, and this is the concept of force.

Considered from a modern point of view, Kepler's introduction of an attractive force as the cause of the fluctuations in speed of the planets is a methodologically justified process since it

[21] "Est siquidem usitatissimum axioma per universam Philosophiam naturalem: Eorum, quae simul et eodem modo fiunt, et easdem ubique dimensiones accipiunt, alterum alterius causam aut utrumque ejusdem causae effectum esse. Ut hic intentio et remissio motus, cum accessu et recessu a centro mundi, in proportione perpetuo coincidit." Kepler, *Astronomia nova,* pars tertia, cap. XXXIII, p. 236.

reduces numerous cases of functional dependences to one single supposed agency. For Kepler himself, of course, such considerations would have been out of the question. For him, the force by which the sun attracts the planets was a physically demonstrated fact. Since force, in his view, has only an immaterial existence, its presence can be recognized only by its effects, which, in turn, underlie the mathematical relation between speed and distance. Moreover, although Kepler was led to the assumption of force by the changes in planetary speed, he still regards force as necessary merely to maintain the planetary motion. Galileo's principle of inertia and Huygens' conception of centripetal forces were, of course, still unknown to Kepler; force was "promotive," but not centrally attractive.[22]

So was Newton's conception of a universal attraction. In his search for possible other phenomena in which the attractive force of the sun may become perceptible or demonstrable, Kepler was thus left with only one possibility: Gilbert's magnetic forces. No wonder that Kepler, when writing his *Tertius interveniens,* is convinced beyond any shade of doubt that his astronomical computations only confirmed his previous assumptions about the importance of magnetic forces. In article No. 51 of this short treatise Kepler asserts emphatically: "The planets are magnets and are driven around by the sun by magnetic force.[23]

Kepler imagined these magnetic forces, emanating from the central body such as the sun, to be like giant arms propelling the planets on their appropriate orbits.[24] He thought it necessary

[22] "Solis corpus est circulariter magneticum et convertitur in suo spacio, transferens orbem virtutis suae, quae non est attractoria sed promotoria." Kepler, *Gesammelte Werke,* ed. Caspar, vol. 15, p. 172, letter to Maestlin, March 5, 1605.

[23] Johannes Kepler, *Tertius interveniens* (Godtfriedt Tampachs, Frankfurt a. M., 1610); *Gesammelte Werke,* ed. Max Caspar and Franz Hammer, vol. 4, p. 192.

[24] Kepler's idea of "magnetic spokes" emanating from the sun and pushing the planets around in their orbits has, curiously enough, been revived in a certain sense in modern cosmological magnetohydrodynamics. See, for instance, Alfvén's hypothesis concerning the transfer of angular momentum from the

that for this purpose the central body itself should be in rotational motion. Thus he predicted the rotation of the sun a short time before the invention of the telescope made it possible to observe this phenomenon. A celestial body, however, Kepler claimed, without a satellite should not possess rotational motion since such would be superfluous. Thus he explains the absence of the rotation (!) of the moon: "Gyratio igitur in Luna, ut supervacua, fuit omissa."[25]

An unmistakable expression of his new conception of force as something corporeal, that is, mechanical, is clearly found in his annotations to the second edition of his *Mysterium cosmographicum* in 1621. Here he says explicitly:

> If you substitute for the word "soul" the word "force," you have the very principle on which the celestial physics of the treatise on Mars etc. is based . . . Formerly I believed that the cause of the planetary motion is a soul, fascinated as I was by the teachings of J. C. Scaliger on the motory intelligences. But when I realized that these motive causes attenuate with the distance from the sun, I came to the conclusion that this force is something corporeal, if not so properly, at least in a certain sense."[26]

Kepler here refers to Julius Caesar Scaliger's book *Exercitationes exotericae,"*[27] which was very popular in Kepler's time and which apparently had a strong influence also on his mind. Kepler's words just quoted announce the birth of the Newtonian concept of force (with which we are so familiar from our study of classical mechanics).

---

sun to the gas out of which the planets were formed; H. Alfvén, *On the origin of the solar system* (Clarendon Press, Oxford, 1955).

[25] Kepler, *Epitome astronomiae copernicanae,* lib. IV, pars. II., in *Gesammelte Werke,* ed. Caspar, vol. 7, p. 319.

[26] "Si pro voce anima vocem vim substituas, habes ipsissimum principium, ex quo Physica Coelestis in Comment. Martis est constituta et lib. IV Epitomes Astr. exculta. Olim enim causam moventem planetas absolute animam esse credebam, quippe imbutus dogmatibus I. C. Scaligeri de motricibus intelligentiis. At cum perpenderem hanc causam motricem debilitari cum distantia a Sole: hinc conclusi vim hanc esse corporeum aliquid, si non proprie, saltem aequivoce." *Joannis Kepleri Opera omnia,* op. I, ed. Frisch (Frankfurt and Erlangen, 1858–1871), p. 176.

[27] Julius Caesar Scaliger, *Exotericarum exercitationum liber quintus decimus de Subtilitate* (F. Morellus, Paris, 1557).

Indeed, in his letter to Fabricius, quoted on p. 83, Kepler asserted that not only does the earth attract the stone, but also the stone attracts the earth. In the introduction to his *Astronomia nova* he proclaims that the moon is attracted by the earth as well as the earth by the moon. The idea of this reciprocity should have divested the concept of force of all animistic ingredients, since force no longer belongs to one single object; it contains already in its definition a necessary relation to a second element. "Der Relationsbegriff ist es, der den Kraftbegriff nötigt, gleichsam aus sich selber herauszutreten und sich in einer reinen mathematischen Proportion zu beugen."[28] Now it became natural to ask for the mathematical formula of the intensity of force. Kepler conjectures first that this intensity is proportional to the reciprocal value of the square of the distance, but soon rejects this suggestion. In his view attractive forces do not spread through space in all directions, as exemplified by light, but only in the plane of the planetary orbits. Thinking still along Peripatetic lines, Kepler assumes that force is proportional to the velocity v of the moving object. From his "second law," that is, from rv = constant, he concludes that the velocity v is proportional to $1/r$, that is, to the reciprocal of the distance between the planet and the sun. He thus asserts that the force is proportional to the reciprocal of the distance. Irrespective of this error, it was Kepler who transformed the concept of force from its Platonic form and interpretation to an essentially relational concept. Thus the concept of force became a basic element of the conceptual apparatus of the seventeenth century.

In conclusion of our discussion of Kepler's decisive contribution to our subject, the summarizing question may be asked: What reasons induced Kepler to introduce the concept of force into the exact sciences? The answer is: three circumstantial reasons and one technical-methodological reason. (1) The Copernican revolution relinquished the Aristotelian doctrine of natural

[28] Ernst Cassirer, *Das Erkenntnisproblem* (Bruno Cassirer, Berlin, 1922), vol. 1, p. 278.

places and of the absolutely heavy and light; (2) the phenomenon of the tides proved to be an incontestable evidence of a direct mechanical influence of the moon (meanwhile conceived as physically similar to the earth) upon the earth; (3) Gilbert's theory gave an illustration of a possible *modus operandi* of that force. We have called these three reasons circumstantial, because they were borne out by the circumstances of the time in which Kepler lived. The decisive reason, however, was the technical-methodological one: his realization of the dependence of planetary velocity upon distance and his desire for a causal explanation.

Kepler's conception of mutual attractive forces exerted by the celestial bodies spread only slowly through Europe in the first half of the seventeenth century. Mersenne, who was one of the first to accept Kepler's ideas, alluded to this conception in his *Synopsis mathematica.* Étienne Pascal and Roberval, in a letter to Pierre Fermat, dated August 16, 1636, write:

> It may be true and it seems very probable that gravity is a mutual attraction or a natural desire ("un désir naturel") of bodies to come together, as is obvious in the case of the iron and the magnet where, if the magnet is arrested, the iron, being free to move, tends to approach it; if the iron, however, is arrested, the magnet will approach it; and if both of them are free to move, they will draw near each other reciprocally, so that in any case the stronger of the two will take the shorter path.[29]

That Kepler's ideas met also with strong opposition can be seen, as late as 1657, in Ismael Bullialdus' *Astronomiae philolaicae fundamenta clarius explicata et asserta,*[30] in which he rejects Kepler's conception of an attractive force exerted by the sun on the planets, calling it a pure imagination not warranted by the facts. In his *Astronomia philolaica* of 1645 Bullialdus had already repudiated Kepler's conclusions, as expounded in the *Astronomia nova* and in the *Epitome,* item by item. Bullialdus does not recognize Kepler's main argument,

[29] Pierre de Fermat, *Oeuvres,* ed. P. Tannery and C. Henry (Gauthier-Villars, Paris, 1894), vol. 2, p. 36.

[30] Ismael Bullialdus (Boulliau), *Astronomia Philolaica, opus novum, in quo motus planetarum per novam ac veram hypothesim demonstrantur* (Paris, 1645); rev. ed., *Astronomia Philolaica clarius asserta* (Paris, 1657).

according to which the existence of a force of attraction can be inferred from the slow motion of the planet at aphelion and its fast motion at perihelion; for Bullialdus this variation of speed is simply a fundamental fact, thus established by the Divine Architect. The sun, although the source of light and heat, is not necessarily also the fount of a force of attraction. Moreover, such a force, if it existed, would be propagated not within a single plane, but would spread through all space; its magnitude would correspondingly decrease with the square of the distance. Bullialdus' main argument, however, is his metaphysical tenet that the principle of action must lie in the agent itself. Just as the fire contains in itself the principle of combustion, and light the principle of illumination, so also the mobile must contain in itself the principle of its motion. The motion of a planet has consequently to have its cause in the planet itself. Needless to say, Bullialdus could not reconcile his revived Aristotelian conceptions with the undeniable fact of the elliptical paths of the planets.

CHAPTER 6

"FORCE" AND THE RISE OF CLASSICAL MECHANICS

Galileo is generally credited with having laid the foundations of classical dynamics, an appraisal that is wholly justified. However, as far as the classical concept of force is concerned, Galileo's contribution may be regarded as complementary to that of Kepler. While Kepler arrived at his concept of force as a result of his far-reaching astronomical investigations, Galileo studied the kinematic aspects of motion, motion originated mainly by a constant force, without, however, delving into the nature of force itself. In fact, in his *Dialogue on the great world systems,* as well as in his *Dialogues concerning two new sciences,* he emphatically rejects any premature conjecture about the "true essence" of force. In the *Dialogues concerning two new sciences* he declares in the person of Salviati:

> The present does not seem to be the proper time to investigate the cause of the acceleration of natural motion concerning which various opinions have been expressed by various philosophers, some explaining it by attraction to the center, others to repulsion between the very small parts of the body, while still others attribute it to a certain stress in the surrounding medium which closes in behind the falling body and drives it from one of its positions to another. Now, all these fantasies, and others too, ought to be examined; but it is not really worth while. At present it is the purpose of our Author merely to investigate and to demonstrate some of the properties of accelerated motion (whatever the cause of this acceleration may be).[1]

[1] Galileo Galilei, *Discorsi e dimostrazioni matematiche intorno a due nuove scienze* (Leyden, 1638); *Dialogues concerning two new sciences,* trans. Henry Crew and Alfonso de Salvio (Northwestern University Press, Evanston, 1946), p. 160.

Galileo's sound instinct revolted here against the attempts of his contemporaries to come to close quarters with the concept of force by purely metaphysical considerations. For Galileo, "force" is first and foremost a physical concept, the exact determination of which was still beyond his power. But Galileo felt certain that his approach, though it was only the beginning, would lead to the desired answer. "And what I consider more important, there have been opened up to this vast and most excellent science, of which my work is merely the beginning, ways and means by which other minds more acute than mine will explore its remote corners." [2]

Galileo's apparently positivistic approach, however, should not suggest that he did not take an interest in the concept of force. One need only read his *Dialogues concerning two new sciences,* and in particular the discussion of the third day, to see that he was grappling with an intuitive notion of force and searching for an exact formulation. He was brought up in the Aristotelian physics which viewed force as a pull or push that moves a body in a direction contrary to that in which it would have moved according to its own natural motion. "For one says that something is moved by force," explains Galileo's teacher Buonamici, "if that which is moved does not receive this force for itself, that is, if it does not in itself possess the propensity by virtue of which it moves, since through this movement it does not perfect itself by reaching the place in which it is conserved." [3] Force, roughly speaking, was an agency that caused unnatural motion and was, so to say, an intruder in the otherwise harmonious system of natural processes. Apart from this conception, Italian naturalistic thought of the sixteenth century employed the term *forza* also for the designation of impetus and similar notions. Take, for instance, Leonardo da Vinci's definition of *forza:*

[2] *Ibid.*, p. 147.

[3] "Vi autem moveri illa dicuntur quandocunque id quod movetur non confert vim, hoc est non habet illo propensionem, quo movetur, quia. s. non perficiatur ex eo motu, locum illum adipiscens in quo conservetur." Giovanni Francesco Buonamici, *De motu* (Apud Sermatellium, Florence, 1591), liber V, cap. 35.

Force I define as an incorporeal agency, an invisible power, which by means of unforeseen external pressure is caused by the movement stored up and diffused within bodies which are withheld and turned aside from their natural use; imparting to these an active life of marvellous power it constrains all created things to change of form and position, and hastens furiously to its desired death, changing as it goes according to circumstances."[4]

Although Leonardo's expression of "bodies withheld and turned aside from their natural use" contains undoubtedly an implication similar to that which was given later by Buonamici, what Leonardo defines is rather impetus, or perhaps what we call today kinetic energy. No doubt, Leonardo surpassed Peripatetic conceptions of motion through his dynamic conception of movement, which for him is not only change of place but the expression of an inherent activity. The vague, ambiguous, and multifarious nomenclature employed indiscriminately to denote force, impetus, momentum, energy, and so forth, is the result of the confluence of Platonic ideas with Aristotelian conceptions. Still with Galileo we find this variety of synonyms for the designation of "force": *forza, potenze, virtù, possanza, momento della potenza,* etc. In discussing the intensity of what we would call the component of force in the direction of an inclined plane, Galileo, groping for the right word, says: "Therefore the impetus, ability, energy, or, one might say, the momentum of descent of the moving body is diminished by the plane upon which it is supported and along which it rolls."[5]

In Galileo's early writings[6] the concept of force is rather frequently employed in relation with the traditional problems of statics, as for the computation of the mechanical advantage of machines, for instance; but it is force as equivalent to weight,

[4] *The notebooks of Leonardo da Vinci,* ed. Edward McCurdy (Braziller, New York, 1956), p. 520.

[5] The Italian original reads: "l'impeto, il talento, l'energia, o vogliamo dire il momento del discendere." Galileo, *Dialogues concerning two new sciences,* p. 174.

[6] Galileo Galilei, *Della scienza meccanica* (published posthumously in Ravenna, 1649; French translation by P. Mersenne, Paris, 1634).

it is force in the Archimedean conception without further analysis. It is this concept of force that is used by Galileo for the principle of equilibrium in machines, according to which force can be decreased at the expense of time.[7]

Weight is for Galileo, at this stage, a natural inclination of a body to come nearer to the center of the world. In his dialogue *De motu,* which represents most clearly his early dynamical conceptions (the so-called Pisan dynamics), Galileo analyzes the concept of weight as follows. By nature, he says, things are so constituted that the heavier bodies rest beneath the lighter bodies. It would be futile, he continues, to assign a cause to such an order; it was pleasing to nature to establish things in this manner: "placuit autem Summae Providentiae in hunc distribuere." And yet, this kind of reasoning, in a typical Peripatetic manner, does not satisfy him; in his search for an explanation of weight he looks for some deeper and more rational reason. "Unless perhaps we might wish to say," he continues,

> that the heavier bodies are nearer to the center than the light ones, because in a way it seems that those things are heavier which, in a smaller space, contain more matter . . . Since then the spaces which are nearer to the center of the world are smaller than those which are more distant from the center, it was in accord with reason that those spaces should be filled with matter which, being of greater weight, would occupy less space.[8]

Impeto becomes his favorite expression for the instantaneous action of a force and is frequently taken for what we call "impulse," as, for example, in his *Discorso intorno alle cose che stanno in su l'acqua:* "Bodies of equal weights and moved with equal velocities have equal forces and moments; equal weights with unequal velocities have moments in proportion to their velocities."[9] Sometimes, however, the term *impeto* denotes in

[7] An idea expressed already in the *Problemata* ascribed to Aristotle.

[8] *Le opere de Galileo Galilei* (national ed., Florence), vol. 1 (1890), p. 252. The English quotation is from E. A. Moody, "Galileo and Avempace," *Journal of the History of Ideas 12,* 168 (1951).

[9] "Che pesi assolutamente eguali, mossi con eguali velocità, sono di force e di momenti eguali nel loro operare. Che pesi assolutamente eguali, ma con-

Galileo the velocity acquired by the mobile in the course of a given time, and in some cases even the distance traversed in that time.

The term used frequently by Galileo in his writings to denote the action of a force is the word "moment" (*momento*). Galileo's conception of this term is complicated by the fact that an originally static concept becomes transferred to a dynamic idea. Originally, moment was conceived as the torque of a force. Thus Giovanni Battista Benedetti, whom, according to Mach[10] and Wohlwill,[11] Galileo seemed to have followed in his early stages, defines moment as the torque, taking into consideration the distance of the line of action of the force from the center of rotation,[12] but conceives this moment as indicative of the motive force (*virtus movens*) exerted on the weight. In his *Della scienza meccanica* Galileo employs the term moment as a purely dynamic concept, recalling perhaps that etymologically "moment" is derived from *movimentum,* that is, "appertaining to motion." He says: "The moment is that impetuosity (*impeto*) to descend which is composed of the ponderosity, the place and other things by which such a propensity (*propensione*) may be caused." [13]

In the same work Galileo seems to justify the use of the term "moment" when he says: "Moment is that force (*virtù, forza, efficia*) with which the mover moves the mobile and with which the mobile resists this motion, which force (*virtù*) depends not

giunti con velocità diseguali, sieno di forza, momento e virtù diseguale, e più potente il più veloce, secondo la proporzione della velocità sua alla velocità dell 'altro." Galileo Galilei, *Discorso intorno alle cose che stanno in su l'acqua o che in quella si muovono* (Florence, 1612); reprinted in *Opere* (national ed.), vol. 4 (1894), p. 68.

[10] Ernst Mach, *The science of mechanics,* trans. T. J. McCormack (Open Court, La Salle, Ill., 1942), pp. 154ff.

[11] Emil Wohlwill, *Galileo und sein Kampf für die Kopernikanische Lehre* (Voss, Hamburg and Leipzig, 1909), vol. 1, p. 115.

[12] Giovanni Battista Benedetti (Benedicti), *Diversarum speculationum mathematicarum et physicarum liber* (Bevilaquae, Turin, 1585).

[13] "Il momento quell' impeto di andare al basso, composto di gravità, posizione e di altro, dal che possa essere tal propensione cagionata." Galileo, "Delle utilità che si traggono dalla scienza mecanica e dai suoi instrumenti," *Della scienza meccanica* (national ed.), vol. II (1891), p. 159.

only upon simple weight, but also upon the velocity of motion and the various inclinations of the spaces in which the motion occurs."

In his *Dialogues concerning two new sciences,* however, the term *momento* or sometimes also the longer expression *momento della potenza* usually denotes "force" in the static sense.[14]

That Galileo, who is often referred to as the first great scientist who replaced the question "why" by "how," was not satisfied with the mere description of free fall and was searching for a more general principle from which the laws of free fall could be deduced, is apparent from his letter to Paolo Sarpi. Galileo writes: "While thinking about the problems of motion, for which I am in want of an absolutely indubitable principle which I can posit as an axiom in order to demonstrate the accidental phenomena observed by me, I arrived at a proposition which has much of the natural and evident in it."[15] In fact, as pointed out by Koyré,[16] Galileo's approach is not that of a positivistic physicist, in the modern sense of the term.

In his investigation of the laws of free fall Galileo disproves the assumption that "immediately after a heavy body starts from rest it acquires a very considerable speed" by pointing out the following argument:

> Imagine a heavy stone held in the air at rest; the support is removed and the stone set free; then since it is heavier than the air it begins to fall, and not with uniform motion but slowly at the beginning and with a continuously accelerated motion. Now since velocity can be increased and diminished without limit, what reason is there to believe that such a moving body starting with infinite slowness, that is, from

[14] Galileo, *Dialogues concerning two new sciences,* pp. 108–111, 120, 121, 131 (national ed., pp. 154–156, 164, 165, 174).

[15] Letter to Paolo Sarpi, Padua, 16 October 1604, "Ripensando circa le cose del moto, nella quali, per dimostrare li accidenti da me osservati, mi mancava principio totalmente indubitabile da poter porlo per assioma, mi son ridoto ad una proposizione la quale ha molto del naturale et dell' evidente." Galileo, *Opere* (national ed.), vol. X, "Carteggio 1574–1642," p. 115.

[16] A. Koyré, *La loi de la chute des corps, Descartes et Galilée* (Études Galiléennes II, Actualités scientifiques et industrielles No. 853; Hermann, Paris, 1939), p. 79 (II–5): "L'épistémologie galiléenne n'est pas positiviste."

rest, immediately acquires a speed of ten degrees rather than one of four, or of two, or of one, or of a half, or of a hundredth; or, indeed, of any of the infinite number of small values? Pray listen. I hardly think you will refuse to grant that the gain of speed of the stone falling from rest follows the same sequence as the diminution and loss of this same speed when, by some impelling force, the stone is thrown to its former elevation.[17]

It is in this connection that Galileo compares muscular force (the impelling force) with the force of gravity. In fact, when the stone, thrown upward, attains its highest position, its momentary state of rest is for Galileo an indication that these two forces, the "impressed impetus" and "the weight of the body," are in equilibrium. Furthermore,

the force (*virtù*) of the projector may be just enough to exactly balance the resistance of gravity so that the body is not lifted at all but merely sustained. When one holds a stone in his hand does he do anything but give it a force impelling (*virtù impellente*) it upwards equal to the power (*facolta*) of gravity drawing it downwards? And do you not continuously impress this force (*virtù*) upon the stone as long as you hold it in the hand? [18]

Obviously, Galileo here comes very near the classical determination of force, could he only have had a clearer conception of mass. Still, he feels instinctively that these considerations will throw some light on the concept of force, or, in his own words:

From these considerations it appears to me that we may obtain a proper solution of the problem discussed by philosophers, namely, what causes the acceleration in the natural motion of heavy bodies.[19]

Although the impressed force of the agent is here still conceived as an impetus that is gradually consumed by the opposing force of gravity, this passage shows that Galileo attempted to understand the force of gravity by relating it to other kinds of forces. Still more striking, in this respect, is Galileo's measurement of the impelling force (*impeto*), which, as we have seen, is

[17] Galileo, *Dialogues concerning two new sciences*, p. 157.
[18] *Ibid.*, p. 159.
[19] *Ibid.*, p. 158.

the component of the gravitating force in the direction of the inclined plane, by a static force. In the same *Dialogues* Salviati says:

> It is clear that the impelling force acting on a body in descent is equal to the resistance or least force (*resistenza o forza minima*) sufficient to hold it at rest. In order to measure this force and resistance (*forza e resistenza*) I propose to use the weight of another body.[20]

In fact, this passage is perhaps the first statement of a unified concept of static and dynamic forces. Galileo, however, does not yet arrive at a mathematical definition of dynamic force, the reason being, as stated, that he still does not possess a clear definition of mass. Yet he already reduces the action of force to a gradual increase of velocity, to the accumulation of increments of speed, an idea that was possible only after he had assumed, at least implicitly, the principle of inertia. In other words, force for Galileo is a continuous sequence of instantaneous impulses that are added to one another. It would have been natural, now, to presuppose the principle of inertia and to conceive force, not as an inherent quality or property of the mobile, but as an impact, an activity from outside, and to conceive it as the cause of the acceleration or increase (more generally, as the change of speed with time) of the moving object. It was Galileo who prepared this conclusion, which forms the basis of the first two laws of motion as formulated by Newton. It was the concept of a force-free motion that was the presupposition of the Newtonian conception of force.

As to the ultimate nature of weight, Galileo made it clear that its essence was hidden from him. The problem, in his view, has to be left as an ultimate unsolved question. His great self-restraint, which for Burtt[21] is an evidence of Galileo's revolutionary greatness, is most clearly expressed in a passage in his

[20] *Ibid.*, p. 175.

[21] E. A. Burtt, *The metaphysical foundations of modern physical science* (Routledge and Kegan Paul, London, 1950; first ed., 1924), p. 93.

Dialogue on the great world systems, in which Salviati says:

I shall say that that which makes the Earth move is a vertue like that by which Mars and Jupiter are moved, and that with which he believes the starry sphere itself does move too. And if he will but inform me what is the mover of one of these bodies, I will be able to tell him what makes the Earth move. Even more, I will undertake to do it, if he can only tell me what moves the parts of the Earth downwards.

Whereupon Simplicius responds:

The cause of this is most manifest, and everyone knows that it is gravity.

And now Galileo answers in the words of Salviati:

You are out, Simplicius; you should say that everyone knows that it is *called* gravity, and I do not question you about the name but about the essence (*essenza*) of the thing. Of this you know not a tittle more than you know the essence of the mover of the stars in gyration, unless it be the name that has been put to the former and made familiar and domestic by the many experiences that we have of it every hour in the day. But we do not really understand what principle of vertue moves a stone downwards any more than we know what moves it upwards, when it is separated from the projector, or what moves the Moon round, except possibly (as I have said) only that name, which more particularly and properly we have assigned to the motion of descent, namely, Gravity. For the cause of circular motion, in more general terms we assign "vertue impressed" (*"virtù impressa"*) and call the same an "intelligence," either *assisting* or *informing;* and to infinite other motions we ascribe *Nature* for their cause.[22]

This passage is of great interest: it not only shows that Galileo related terrestrial gravity to the motive forces of the planets, unless he only wanted to express their equally mysterious character, but also indicates that he did not take Gilbert's theory of magnetic forces too seriously, although he discusses this theory in great detail at the end of the third day in the same *Dialogue.*[23]

With the recognition of the law of inertia, formulated already

[22] Galileo Galilei, *Dialogue on the great world systems,* in the Salusbury translation, ed. Giorgio de Santillana (University of Chicago Press, Chicago, 1953), p. 250.

[23] Galileo Galilei, *Dialogo sopra i due massimi sistemi del mondo,* in *Opere* (national ed.), vol. 7 (1897), p. 260.

as early as 1585 by Benedetti as part of his impetus theory and presupposed by Galileo in his analysis of accelerated motion, or with the acceptance of the principle of conservation of motion, as announced by Isaac Beeckmann and René Descartes,[24] two alternative possibilities presented themselves: either to conceive force as the cause of change of motion, or to abolish the notion of force altogether. In any case, velocity as such could no longer be considered an indication of the existence of force or of its measure. How far Beeckmann approached Newton's qualitative definition of force can be seen from his statement: "Bodies, once in motion, will never rest unless impeded (by an external agency)."[25] Beeckmann, however, did not conceive such impediment as a force, a step that would have been a partial enunciation of Newton's qualitative definition of force. He was apparently too much influenced by the view of Descartes, who chose the second alternative: the rejection of the existence of force altogether.

Although Descartes, particularly in his earlier writings, refers to forces of attraction exerted by the earth on a falling object and explains the accelerated motion of such an object by the cumulative action of the force,[26] he eventually conceived "force" as merely a fictitious appearance. His absolute dichotomy of existence into pure matter and pure spirit seemed to him incompatible with the assumption of force in matter or exerted by matter, since force, in his view, is still a somewhat psychic notion. Matter has to be divested of all spiritual constituents, of all inherent forms or tendencies. Only extension and eternal motion are its characteristics. Consequently, in his letter to Mer-

[24] *Oeuvres de Descartes,* ed. Charles Adam and Paul Tannery (Cerf, Paris), vol. 10 (1908), Descartes-Beeckmann (1618–1619), pp. 60, 225.

[25] "Mota semel nunquam quiescunt, nisi impediantur." *Ibid.,* p. 60, note f. "Quod semel movetur semper . . . movetur dum ab extrinseco impediatur." *Ibid.,* p. 225, note b. (Cogitationes privatae.)

[26] "Lapis cadens in vacuo cur semper celerius cadat: Moventur res deorsum ad centrum terrae, vacuo intermedio spatio existente, hoc pacto; primo momento, tantum spacium conficit, quantum per terrae tractionem fieri potest. Secundo, in hoc motu perseverando superadditur motus novus tractionis, ita ut duplex spacium secundo momento peragretur." *Ibid.,* p. 58.

senne[27] dated November 13, 1629, he discusses free fall without any reference to attractive forces. In his endeavor to eliminate the concept of force from his system, he employs here the medieval concept of impetus, but resorts later to his theory of vortices, which for him were of a purely kinematic character. The increasing experience with forces other than gravity, the new, speedily accumulating body of information in pneumatics and hydrostatics, forces that contended with inertia, while gravity was totally irrelevant to them, forced the concept of inertia into the foreground of the theoretical considerations of Descartes, Pierre Gassendi, and Giovanni Battista Baliani. On grounds of the principle of inertia, Descartes thought it might be possible to eliminate force as a separate physical concept. All physical phenomena, he contended, are to be deduced from only two fundamental kinematic assumptions: the law of the conservation of quantity of motion — which for him was not a corollary of the principle of inertia but its real physical content — and his theory of swirling ethereal vortices. For, rejecting any possibility of an action at a distance, Descartes constructs his vortex theory to account for the remote heavenly motions. To assume some action at a distance for their explanation would be tantamount, he claims, to endowing material particles with knowledge and making them truly divine, "as if they could be aware, without intermediation, of what happens in places far away from them." [28]

The concept of force, in Descartes's view, had no place in his physics, which was to employ exclusively mathematical conceptions. "I do not accept," he says in the *Principles of Philosophy*, "or desire any other principle in Physics than in Geometry or abstract Mathematics, because all the phenomena of nature may

[27] *Oeuvres de Descartes,* vol. 1, p. 71.

[28] "Nam ad hoc intelligendum necesse est, non modo supponere singulas materiae particulas esse animatas, et quidem pluribus animabus diversis, quae se mutuo non impediant, sed etiam istas earum animas esse cogitativas, et plane divinas, ut possint cognoscere quid fiat in illis locis longe a se distantibus, sine ullo internuntio, et ibi etiam vires suas exercere." *Oeuvres de Descartes,* vol. 4 (1901), Correspondance: Juillet 1643–Avril 1647, letter to Mersenne, April 20, 1646, p. 401.

be explained by their means, and a sure demonstration can be given of them." [29]

A geometrization of physics — this was Descartes's program before classical mechanics was born. It was a program too daring and too difficult, even for such an intellectual giant as Descartes. To start with, he had to augment his mathematical vocabulary with the concept of impenetrable extension (of solid matter), whose nature as a mathematical concept is most questionable. Furthermore, Descartes's theory of vortices, which for him was a system of essentially kinematic propositions, employed concepts of pressure and similar notions that were dynamic concepts in disguise.

How, then, did Descartes explain the phenomenon of gravity?

> But I desire now that you consider what the gravity of this earth is, that is to say, the force which unites all its parts, and which makes them all tend toward its center, everyone more or less according as they are more or less large and solid; which is nothing else and consists only in this, that the parts of the small heaven which surrounds it, turning much more swiftly than its own do around its center, tend also with much more force to withdraw themselves from it, and consequently push them back there.[30]

Descartes tries to show that the more voluminous particles, constituting the vortex in the center of which the earth is situated, are constantly moving toward the circumference of the vortex, and this owing to their greater bulk; terrestrial matter, cohering in great lumps, cannot yield so easily to the outward pressure as this ethereal substance; this latter, however, cannot reach the outer regions of the vortex unless other particles are shifted down, exchanging their place with that of those. Gravity, in other words, is but the downward motion of terrestrial matter, some kind of antiperistatic turbulence in the plenum. The earthly bodies are pressed toward the center of the vortex, namely, the earth, by the buoyant ethereal elements of the vortex. Gravity,

[29] Descartes, *Selections*, ed. R. M. Eaton (Scribner, New York, 1927), p. xxiii.

[30] *Ibid.*, p. 339.

therefore, is not a tendency inherent in matter, but a repulsion or reaction exerted by the ethereal particles which recede from the center of the vortex. The stone falls to the earth, because it has to give room for the ethereal particles which on account of their circular motion depart from the surface of the earth. If this speedily rotating ethereal matter did not surround the earth, or if the earth were assumed to be plunged into the void, all the bodies that are not securely attached to the ground would be flung off as a result of the circular motion to which they are constantly subjected. Terrestrial bodies would thus appear to be light and not heavy.[31] It is the vast number of gross ethereal particles that constrain the terrestrial bodies to the surface of the earth, much as lead shot drives pieces of wood toward the center in a rapidly rotating spherical vessel, an experiment to which Descartes refers in a letter to Mersenne.

Better known is the following experiment, frequently used as a lecture demonstration on centrifugal force (Fig. 2). Two glass

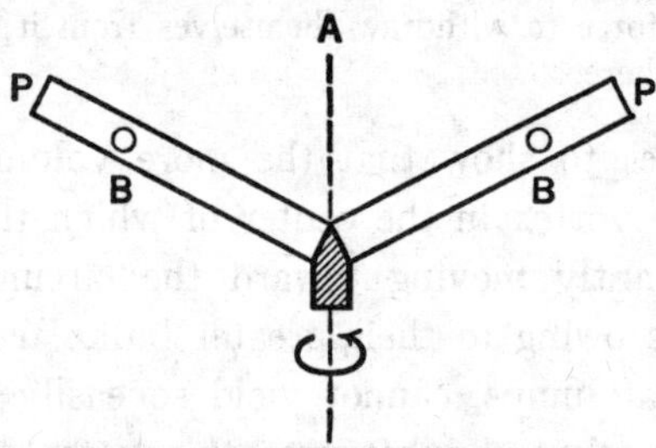

Fig. 2. An experiment to demonstrate Descartes's vortex theory of gravitation.

cylinders, each containing water and a wooden ball *B*, are attached, slightly inclined above the horizontal, to a common axis of rotation *A*. When the system is at rest, the wooden balls will occupy the highest possible places *P* in the glass cylinders; but when the system spins with a certain angular speed, the balls

[31] Descartes, *Oeuvres*, vol. 8 (1905), p. 212.

will be forced toward the center (or axis of rotation). The water in the rotating glass cylinders corresponds to Descartes's vortical medium of ethereal particles and the wooden balls to terrestrial objects. (The resulting centripetal impulsion, expressed in modern terms of classical mechanics, is of course a buoyancy effect in a medium that is subjected to a field of radial inertial forces, the so-called centrifugal force.)

Although Descartes's theory of gravitation was perhaps the first conceptual scheme to unite terrestrial gravitation with the cosmic motion of planets moving in their solar vortex, it did not enjoy much favorable acceptance, at least outside France. Although Cartesian physics as a whole was hailed as an ingenious edifice of thought, its theory of gravitation was considered by the Cartesians themselves, even prior to the advent of Newton's theory of gravitation, as its weakest point. Nevertheless, the Newtonian theory of gravitation as an action at a distance could not easily supersede the Cartesian conceptions. Even as late as 1730, when the French Academy of Sciences proposed a problem on the cause of the ellipticity of the planetary orbits,[32] Jean Bernoulli submitted a solution that was based entirely on the Cartesian conception of vortices. Forces did not enter the discussion because, according to Bernoulli, Descartes's approach explains the nature of weight much better than Newton's attractive virtue. "Les Tourbillons se présentent si naturellement à l'esprit," he says, "qu'on ne sauroit presque se dispenser de les admettre."[33] Although many attempts were made to improve upon certain shortcomings of the vortex theory, as for example by Pierre Varignon[34] or by Joseph Saurin,[35] who tried to invalidate Huygens's

[32] The problem reads: "Quelle est la cause de la figure elliptique des Orbites des Planètes et pourquoy le grand axe de ces Ellipses change de position, ou ce qui revient au même: Pourquoy leur Aphélie, ou leur Apogée répond successivement a differens points du Ciel?"

[33] Jean Bernoulli, *Nouvelles pensées sur le système de M. Descartes* (1730), in *Opera omnia* (Bousquet, Lausanne and Geneva, 1742), vol. 3, p. 142.

[34] Pierre Varignon, *Nouvelles conjectures sur la pesanteur* (Paris, 1691).

[35] Joseph Saurin, "Examen d'une difficulté considérable proposée par Mr. Huyghens contre le système cartésien sur la cause de La pesanteur," *Journal des Sçavans* (1703).

objections[36] that on the grounds of the vortex theory the gravitational motion, that is, the centripetal impulsion, should be directed toward the axis and not to the center of the vortex, and that the vortical medium must have greater density than any terrestrial substance, Descartes's doctrine was regarded as an *ad hoc* theory which, contrary to its proclamation, had to smuggle in the concept of force under the cover of pressure or in another form.

In the first half of the seventeenth century many astronomers, in accord with Kepler's early hypotheses, still believed in a rotational force of magnetic nature exerted by the spinning central body and dragging the planets or satellites along their orbits.

But toward the middle of the century the existence of some kind of central force, directed from the sun to the planets, was generally accepted. But here, as always in science, the solution of one question raises a new problem: Why do not the planets, under the action of this central force, fall into the sun? Various solutions were offered. Thus Giovanni Alphonso Borelli, in his *Theoricae Medicorum Planetorum ex causis physicis deductae,* based on observations of the satellites of Jupiter over a number of years, conceives the planetary orbits as certain equilibrium positions of opposing forces, speaking, in a rather vague fashion, about material ties that bind the planet with its central body, ethereal currents that carry the planet around the sun, and a natural tendency of the planets to approach the sun, which are all counteracted by a certain propensity of the planets to reach the circumference of the planetary system.[37] At times, the material ties connecting the planet to the sun are spoken of as magnetic forces of constant magnitude. The progressive — we would

[36] Huygens, in his *Discours de la cause de la pesanteur* (Leyden, 1690), p. 133, wrote on Descartes's experiment: "Mais ce qui arrive icy n'est nullement propre à représenter l'effet de la pesanteur; puis qu'on devroit conclure de cette expérience, que les corps, qui contienent le moins de matière, sont ceux qui pesent le plus, ce qui est contraire à ce qui s'observe dans la véritable pesanteur."

[37] Giovanni Alphonso Borelli, *Theoricae Medicorum Planetarum ex causis physicis deductae* (Florence, 1666), p. 48.

call it today the tangential — motion along the orbit, however, is accounted for by a moving force, inherent in the planet and decreasing in magnitude with increasing distance from the sun. It is obvious that the assumption of an "inherent" force in the moving body, which nevertheless is related to the distance from the center, is a serious conceptual weakness, if not a logical inconsistency, in Borelli's explanation, even if this relation is conceived as accidental. Borelli shows that a planet, if moving in a path too far from the sun, approaches the sun, since its motion of progression, reduced at that distance, gives way to the tendency of approach. If then, however, the planet draws too near to the sun, this motion exceeds the centripetal tendency and the planet recedes from the sun. Not having the necessary mathematical apparatus at his disposal, Borelli was unable to prove that the resulting planetary orbits are elliptical in shape. In fact, almost all his considerations were only of a qualitative character. Nevertheless, his work on the satellites of Jupiter contributed decisively to the rejection of rotational magnetic forces as the cause of planetary motions and laid the foundations for future gravitational theories. Yet, the mathematical formula for the centrifugal impulsion as a function of the orbital speed was yet to be found.

As is well known, it was Christian Huygens who solved this problem in his *Horologium oscillatorium*,[38] in which the subject of uniform circular motion is treated only incidentally, and without the necessary geometrical demonstrations. Huygens was already in possession of the mathematical expression of the so-called centrifugal force in 1669, as his famous letter to Henry Oldenburg, the secretary of the Royal Society of London, indicates. In this letter, dated September 4, 1669, Huygens asks Oldenburg to keep in the custody of the registers of the Royal Society some anagrams "for avoiding disputes, and for rendering

[38] Christian Huygens, *Horologium oscillatorium sive de motu pendulorum ad horologia aptato demonstrationes geometricae* (Paris, 1673), reprinted in *Oeuvres complètes,* ed. Société Hollandaise des Sciences, vol. 18 (Martinus Nijhoff, The Hague, 1934), pp. 69ff.

to each individual that which is rightly his in the invention of new things."[39] One of the anagrams read as follows:

a	b	c	d	e	f	g	h	i	l	m	n	o	p	q	r	s	t	u	x	y	z
3	0	6	0	7	1	0	0	5	1	3	2	3	2	0	6	3	4	4	1		
9	0	1	3	5	1	2	0	6	1	5	2	0	0	2	4	0	6	5	0,[40]		

and with slight changes is transliterated in Appendix III to *De vi centrifuga,*[41] Theorema V, as: "Si mobile in circumferentia circuli feratur ea celeritate, quam acquirit cadendo ex altitudine, quae sit quartae parti diametri aequalis; habebit vim centrifugam suae gravitati aequalem; hoc est, eadem vi funem quo in centro detinetur intendet, atque cum ex eo suspensum est."[42] In English: "If a body moves in circular motion with the speed it would gain in falling from a height equal to the fourth part of the diameter, then the centrifugal force will be equal to the pull of gravity upon it, i.e., it stretches the string just as much as if it were suspended by it."

In our modern notation the centrifugal force, as demonstrated in full in Huygens's posthumous *De vi centrifuga,* is given by the formula $F = mv^2/r$, in which v denotes the velocity and m the mass of the particle, and r the radius of its circular path. If the particle were to fall through a distance $s = r/2$, that is, a quarter of the diameter, its velocity would be $v = \sqrt{2gs} = \sqrt{gr}$ and its centrifugal force consequently mg, that is, equal to its weight. This shows the correctness, in terms of Newtonian mechanics, of Huygens's assertion for the particular case mentioned on the assumption of the validity of the general formula for F.

Let us now expound Huygens's geometrical proof of this particular case as given in his own writings,[43] but in our modern mathematical notation. His proof, of course, is not based on the foregoing formula, to which it is logically prior. We think it worth

[39] Cf. Henry Crew, *The rise of modern physics* (Williams and Wilkins, Baltimore, 1935), p. 123.

[40] Christian Huygens, *Oeuvres complètes,* vol. 6 (1895), p. 487.

[41] Huygens, *De motu et vi centrifuga, Opuscula postuma* (Leiden, 1703).

[42] Huygens, *Oeuvres complètes,* vol. 16 (1929), p. 316.

[43] *Ibid.,* p. 275.

while to review his proof, because it affords a good illustration of the use of the concept of *conatus,* frequently employed in the seventeenth century to signify the action of a force, or tendency, during a short constant interval of time. The Latin word *conatus,* meaning originally "effort," "exertion," or "impulse," was used already by Cicero in the sense of natural inclination, when he said in his *De natura deorum:* "But nature has also bestowed upon the beasts both sensation and desire, the one to arouse in them the impulse (*conatum*) to appropriate their natural food, the other to enable them to distinguish things harmful from things wholesome."[44] In the seventeenth century the concept of *conatus* became, especially with Huygens, a quantitative, in fact the first quantitative, expression for force, while Spinoza and Hobbes employed the concept chiefly for their geometrized psychological theories of emotions and volitions.

For Huygens the *conatus* of a moving body was measured by the distance that it covered, under the action of a given force, within a unit interval of time. If we ignore for a moment the possibility of a force-free motion of a body, it can be shown easily that Huygens's idea of measuring force by that distance is a sound idea, from the point of view of classical mechanics. Anticipating Newton's second law of motion, we know that the force is proportional to the acceleration of the moving body. According to Galileo's formula $s = \frac{1}{2}gt^2$, the distance traversed within unit time is also proportional to this acceleration; consequently, the force is proportional to this distance. The intensity of the centrifugal force, says Huygens, can therefore be measured by the distance through which the body is deflected, within a small unit of time, from the tangential direction in which it would move in the absence of this force. Let us now assume that a body moves on a circle *HBF* (Fig. 3) with center *A* with the speed that it

[44] "Dedit autem eadem natura beluis et sensum et appetitum, ut altero conatum haberent ad naturales pastus capessendos, altero secernerent pestifera a salutaribus." Cicero, *De natura deorum,* trans. H. Rackham (Loeb Classical Library; Harvard University Press, Cambridge, 1951), p. 238.

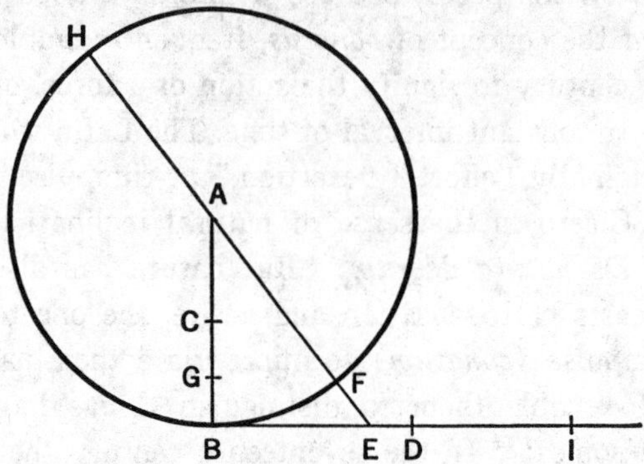

Fig. 3. Huygens's method of measuring centrifugal force.

would attain after falling through CB, that is, a quarter of the diameter FH. This speed is clearly $v = \sqrt{rg}$, if r denotes the radius AB of the circle. The time it takes to fall through CB is, according to the above-mentioned formula of Galileo, $\sqrt{(2CB/g)}$. If the body were now suddenly released from its circular motion at the point B, it would proceed along the tangent BI with its constant speed v, reaching, after the time $\sqrt{(2CB/g)}$, a point D at a distance $\sqrt{(2CB/g)}\sqrt{rg}$, or r, since $CB = r/2$. Let us now consider a small fraction BE of BD and draw the line $EFAH$. The point G on CB is chosen to satisfy the equation $2r \cdot CG = BE^2$. How much time T does it take for the body to fall through CG? Again, $T = \sqrt{(2CG/g)} = \sqrt{(BE^2/rg)} = BE/\sqrt{gr}$ or $BE = \sqrt{(rg)}T = vT$. But since BE is nearly equal to the arc BF, we have $BF = vT$, which shows that it takes the same time to fall through CG as it takes to move along the circle from B to F. Now, CG measures the *conatus* of gravity for free fall, whereas FE measures the *conatus* of the centrifugal force. But $FE \cdot HE$ is approximately equal to $2r \cdot FE$, which again is approximately equal to BE^2, and, by our choice of G, $BE^2 = 2r \cdot CG$; hence it follows that $CG = FE$, that is, the *conatus* of gravity is equal to that of the centrifugal force. Q.E.D.

It is obvious that centrifugal force is conceived by Huygens as a real force, on the same footing as other forces known in his time. The idea of regarding it as a fictional or inertial force is, of course, of a much later date.

Some remarks written for his treatise *De vi centrifuga* as early as in 1659[45] and some passages in his *Horologium oscillatorium*[46] clearly indicate that Huygens, while studying the properties of centrifugal force, realized the possibility that this force may counterbalance the gravitational force exerted by the sun on the planets and thus keep them in their proper orbits.

As to the nature of centrifugal forces or gravitational forces, Huygens wrote very little prior to 1690. In his *Horologium oscillatorium* he adopts a typical Galilean position of demonstrating his mathematical conclusions without discussing any more general (or metaphysical) questions beyond the factual data. This does not imply that he was not interested in the nature of gravity. In fact, when the Parisian Academy of Sciences planned a discussion on the nature of gravity, in 1668, Huygens defended a vortex theory, similar to that advanced by Descartes. And ever thereafter he was working on a treatise on the cause of gravity which he published, however, only in 1690, five years before his death, at a time when Newton has already taken up this matter from a more general point of view and employed it most successfully. It is also well known that Huygens during his first and last visit to the Royal Society of London, in the summer of 1689, spoke in the presence of Newton about his own vortical theory of gravity, while Newton, ironically, discussed the phenomenon of double refraction of light in Iceland spar.[47] Most probably, it

[45] October 21, 1659, Appendix I to *De vicentrifuga*. See *Oeuvres complètes,* vol. 16, pp. 303–304, and also vol. 17, p. 276.

[46] Newton, in the Scholium to proposition IV, book I, of his *Philosophiae naturalis principia mathematica* admits himself: "Mr. Huygens, in his excellent book De horologio oscillatorio, has compared the force of gravity with the centrifugal forces of revolving bodies." See *Sir Isaac Newton's mathematical principles of natural philosophy,* trans. Andrew Motte, rev. Florian Cajori (University of California Press, Berkeley, 1934), pp. 46–47.

[47] Joseph Edleston, *Correspondence of Sir Isaac Newton and Professor Cotes* (London, 1850), p. xxxi.

was this incident that induced Huygens to publish, finally, his *Discours de la cause de la pesanteur.*[48] It may be of some interest today for us, who are familiar with the principles of general relativity and look upon Newton's theory of gravitation as only a good approximation, to learn something about Huygens's opposition.

Huygens was just one of the numerous scientists on the continent who could not reconcile themselves to the conception of force as an action at a distance. For him, as for them, such a conception meant to surrender scientific reasoning to occult qualities. Reared on Cartesian philosophy, he begins his *Discourse* with the following remarks:

> In order to find an intelligible explanation of gravity, one has to find out how it is possible . . . that bodies, supposed as being divested of all qualities and inclinations of mutual approach and as having merely different sizes, shapes and motions, can tend toward a common center and there accumulate; which is the most common and the principal phenomenon of what we call gravity.[49]

Huygens then illustrates his theory of gravity by an experiment that he performed for this special purpose and that he considers remarkable because it exhibits visually his principle of gravity.[50] He placed small particles of sealing wax in a closed vessel containing water and rotated the vessel. The particles of sealing wax, being slightly heavier than the water, moved to the outer rim of the bottom. When the rotational motion of the

[48] Huygens, *Discours de la cause de la pesanteur* (Leiden, 1690), an appendix to his *Traité de la lumière.*

[49] "Pour trouver une cause intelligible de la Pesanteur, il faut voir comment il se peut faire, en ne supposant dans la nature que des corps qui soient faits d'une mesme matiere, dans lesquels on ne considere aucune qualité ni aucune inclination à s'approcher les uns des autres, mais seulement des différentes grandeurs, figures, et mouvements; comment, disje il se peut faire que plusieurs pourtant de ces corps tendent directement vers un mesme centre, et s'y tienent assemblez à l'entour; qui est le plus ordinaire et le principal phenomene de ce que nous appellons pesanteur." *Ibid.*, p. 129.

[50] "L'on peut voir cet effet par une expérience que j'ay faite exprès pour cela, qui mérite bien d'estre remarquée, parce qu'elle fait voir à l'oeil une image de la pesanteur." *Ibid.*, p. 132.

vessel was arrested, but the water still continued to rotate, these particles, touching the bottom and being therefore more rapidly arrested, flocked toward the center of the bottom. This centripetal motion, Huygens proclaims, is an exact reproduction of the mechanism of gravity. After a discussion of the various motions of the fluid matter that surrounds the earth, somehow similar to Descartes's ethereal particles, he identifies gravity with "the effort that makes this fluid matter remove from the center and pushes in its stead those bodies which cannot follow this motion.[51]

To account for Galileo's "naturally accelerated motion" of freely falling bodies and for the fact — a most interesting observation by Huygens — that gravity cannot be screened off[52] by any interposed objects, he assumes that the fluid matter, supposed to be of very fine size and thus capable of penetrating easily through the pores of ordinary matter, imparts to falling bodies continually successive impacts by contact. Thus Huygens thought that he could explain gravitation as a *vis a tergo,* exclusive of any action at a distance. The vortex theories of Descartes and Huygens became the prototypes of all future post-Newtonian mechanistic theories of gravitation, and even had, at least to a limited extent, as we shall see in the following chapters, a not inconsiderable influence on Newton himself.

[51] "C'est donc en cela que consiste vraisemblablement la pesanteur des corps: laquelle on peut dire, que c'est l'effort que fait la matière fluide, qui tourne circulairement autour du centre de la Terre en tous sens, à s'éloigner de ce centre, et à pousser en sa place les corps qui ne suivent pas ce mouvement." *Ibid.,* p. 137.

[52] "Ainsi donc j'ay expliqué, par une Hypothèse qui n'a rien d'impossible, pourquoy les corps terrestres tendent au centre; pourquoy l'action de la gravité ne peut estre empêchée par l'interposition d'aucun corps de ceux que nous connoissons; pourquoy les parties de dedans de chaque corps contribuent toutes a la pesanteur; et pourquoy en fin les corps en tombant augmentent continuellement leur vitesse, et cela dans la raison des temps. Qui sont les propriétéz de la pesanteur qu'on avoit remarquées jusqu'a présent." *Ibid.,* p. 145.

CHAPTER 7

THE NEWTONIAN CONCEPT OF FORCE

Much has been written on Newton's concept of gravitation but next to nothing on his concept of force in general. His ideas on force as such are expounded in the introductory chapter of his *Philosophiae naturalis principia mathematica,* submitted to the Royal Society in 1686 and published in 1687. In fact, the almost complete lack of comment on this subject is rather surprising, not only because the *Principia* is generally regarded, and rightfully so, as the first systematic deductive exposition of classical mechanics, but also because Newton's concept of force, as we shall see, is by no means a simple one.

Newton's concept of force is intimately related, historically as well as methodologically, to his profound study of gravitation. The relation is historical, because his exposition of the fundamental ideas of mechanics was intended to serve, first and foremost, as an axiomatic foundation for his mathematical deductions of gravitational phenomena in the solar system. We know from a report by Edleston[1] that Newton's treatise *De motu,* which is the title he gave originally to the first book of the *Principia,* is part of the lectures delivered by Newton as Lucasian Professor which commenced in October 1684, at a time when he became interested again in the problem of gravitation. In fact, Hooke's letter to Newton of November 24, 1679, in which Hooke

[1] Sir David Brewster, *Memoirs of the life, writings, and discoveries of Sir Isaac Newton* (Edinburgh, 1855), vol. 1, p. 300.

invited him to resume his scientific correspondence with the Royal Society, and the subsequent discussion on the determination of the trajectory of a falling body *in hypothesi terrae motae,* had already renewed Newton's interest in the problem of gravitation.[2]

Newton's general considerations about force are methodologically related to his study of gravitation because the problem of a dynamical explanation of planetary motions to account for Kepler's three laws was the question of the hour. Robert Hooke, Edmund Halley, and many other of Newton's contemporaries were busily engaged in this query and in the investigation of terrestrial gravity. John Wallis, Sir Christopher Wren, and Hooke also advanced numerous suggestions for the solution of these questions. In particular, it was generally agreed that Kepler's laws seemed almost to imply that the force exerted by the sun on the planets decreased with distance. Indeed this was implied by Kepler himself[3] and an inverse-square law had been suggested by Bullialdus.[4] We also know that early in 1684 Hooke mentioned to Wren and Halley, probably in one of their numerous coffeehouse meetings, that the planetary motions could be deduced mathematically from an inverse-square law. Hooke claimed to be in possession of such a mathematical proof, whereupon Wren, expressing his doubts, promised Hooke a book costing forty shillings if, within the following two months, he would produce such a proof. Although Hooke, as is well known, failed to solve the mathematical problem, he and his friends, nevertheless, believed in the validity of the inverse-square law. With respect to terrestrial gravity the situation was similar. Hooke performed a number of experiments at Westminster Abbey, Saint Paul's

[2] Cf. Jean Pelseneer, "Une lettre inédite de Newton," *Isis 12,* 237 (1929). For the early history of the problem of the falling body see Alexandre Koyré, "A documentary history of fall from Kepler to Newton (Gravium naturaliter cadentium in hypothesi terrae motae)," *Transactions of the American Philosophical Society* [n.s.] *45,* part 4 (1955). See also Newton's letter to Halley, referred to in reference 30.

[3] See p. 90, reference 26.

[4] See p. 93.

Church, and the "Column on Fish Street Hill," to detect a diminution of gravity with height. So firmly convinced was he that weight decreases with height that he considered the negative outcome of his experiments merely inconclusive, due to the insufficient precision of his instruments.[5]

When Newton identified terrestrial gravity with astronomical attraction and attempted to corroborate his theory by the famous computations regarding lunar motion, he realized that the weight of a terrestrial object is a function of its distance from the center of the earth, and that the acceleration of a falling body is a function of its position in space. By the assumption that some characteristic property of the body must be invariant with respect to its position, Newton was led to a clear distinction between weight and mass, or what he called "quantity of matter," a notion that had been conceived already by Kepler, Gilbert, and Galileo, but before Newton had never been consciously and explicitly recognized as a basic concept in mechanics. When searching for a mathematical determination of this new concept, it seems probable that Newton recalled Robert Boyle's experiments on compressing gases, in which the quantity of the gas under varying pressure was determined by the product of its volume and its density. Possibly it was this analogy that led Newton to his definition of "quantity of matter" as the product of bulk and density.[6] Once the concept of mass became a fundamental notion of the conceptual system, the definition of momentum, or, in Newton's words, "quantity of motion," and of force as determined by the change of momentum was no longer a difficult task, if Galileo's principle of inertia and Huygen's preparatory work in this line were duly considered.

[5] Robert Hooke, *Posthumous works,* ed. Richard Waller (London, 1705), pp. 176–186.

[6] "Definition I: The quantity of matter is the measure of the same, arising from its density and bulk conjointly." *Sir Isaac Newton's mathematical principles of natural philosophy,* trans. Andrew Motte, rev. Florian Cajori (University of California Press, Berkeley, 1934), p. 1. (Quoted hereafter as *Principles.*)

It is obvious that this trend of thought, which suggests that Newton's concept of force came from his preoccupation with problems of celestial dynamics, is only suggested here as a hypothesis. The *Principia* is written *more geometrico* as a hypothetical deductive system. About its psychological genesis, or, in other words, about the intellectual development of its authorship, we know too little to reach incontrovertible conclusions. In any case, it is clear that the order of presentation in this book does not necessarily reflect the original sequence of ideas in the mind of its author. Yet, a slight confirmation of the ideas suggested above may be read out of Newton's preface to the first edition: "I offer this work as the mathematical principles of philosophy, for the whole burden of philosophy seems to consist in this — from the phenomena of motions to investigate the forces of nature, and then from these forces to demonstrate the other phenomena." [7]

Let us now turn to a detailed discussion of Newton's definition of force. The term "force" (*vis*) appears for the first time in Newton's magnum opus in Definition III: "Materiae vis insita est potentia resistendi, qua corpus unumquodque, quantum in se est, perseverat in statu suo vel quiescendi vel movendi uniformiter in directum." The English translation reads: "The *vis insita*, or innate force of matter, is a power of resisting by which every body, as much as in it lies, continues in its present state, whether it be of rest, or of moving uniformly forwards in a right line." [8] The inert nature of matter is here conceived as a force of inactivity. Inertia, in Newton's opinion, is some kind of force that is inherent (*insita*) in matter and latent as long as no other force, impressed upon the body, "endeavors to change its condition." Moreover, it may be considered both as resistance and as impulse. Newton himself admits: "It is resistance so far as the body, for maintaining its present state, opposes the force impressed; it is impulse so far as the body, by not easily giving way

[7] *Ibid.*, p. xvii.
[8] *Ibid.*, p. 2.

to the impressed force of another, endeavors to change the state of that other."[9] It is resistance if the body is at rest, impulse if it is in motion. But this distinction can be only an apparent one, since motion and rest, as Newton himself claims at the end of this section, are only relatively distinguished.

Clearly, in this definition, force is not conceived as a cause of motion or acceleration. How, then, was it possible for Newton to call the quality of inertia a force? The answer to this question is evident if we regard Definition III as a concession to pre-Galilean mechanics. As we have seen, Peripatetic mechanics conceived force (*dynamis*) as having a dual nature, an active one insofar as it affects other objects, and a passive one as susceptible of external changes. It is this passive connotation that supplies the answer. In fact, Thomas Le Seur and Franciscus Jacquier, who republished an edition of the *Principia* in 1760, comment on Definition III as follows: "Vis duplex est, activa et passiva; Activa est potentia motum efficiendi; Passiva potentia recipiendi vel amittendi."[10] As far as the inherence of force is referred to the body in motion, Newton's definition is reminiscent also of Descartes's conception of force. In fact, if we compare the first part of Descartes's First Law of Nature with Newton's Definition III we see a striking similarity. Descartes wrote: "Prima lex naturae: quod unaquaeque res, quantum in se est, semper in eodem statu perseveret."[11] There is, however, a slight difference in the meaning of "quantum in se est": for Descartes it meant the spatial extension of the body, in conformity with his identification of matter with space; for Newton it is matter itself. For Newton, inertia is proportional to the quantity of matter possessed by the body.

In contrast to "innate force," Definition IV of Newton's *Principia* defines "impressed force" thus: "An impressed force is an

[9] *Ibid.*

[10] *Philosophiae naturalis principia mathematica,* ed. Thomas Le Seur and Franciscus Jacquier (Glasgow, 1760), p. 4, note d.

[11] *Oeuvres de Descartes,* ed. Charles Adam and Paul Tannery (Paris, 1905), vol. 8, *Principia philosophiae,* pars secunda, cap. XXXVII, p. 62.

action exerted upon a body, in order to change its state, either of rest, or of uniform motion in a right line." [12] Or in the original Latin: "Vis impressa est actio in corpus exercita, ad mutandum ejus statum vel quiescendi vel movendi uniformiter in directum." Three aspects distinguish impressed force from innate force: first of all it is pure action, transeunt in character; second, it no longer remains in the body once the action is over; finally, whereas inertia, in Newton's view, is a universal force of matter, innate, but not further reducible, impressed force may have different origins, as "percussion, pressure, or centripetal force." What Newton had in mind when he said, "Centripetal force originates impressed force," seems to be the following: centripetal force, or central force, is an ultimately irreducible element of the conceptual apparatus, an element that manifests itself by the impressed force exerted upon a body and measurable by the change of its momentum. Newton's definition of impressed force as changing the state of rest or uniform motion of a body is closely related to his metaphysical principle of causality. Since every change must have its cause, the change of motion is an effect and the impressed force its cause. When Newton speaks of the ephemeral character of this impressed force, he seems to have had in the back of his mind the old scholastic dictum: *Cessante causa cessat effectus.*

The fifth Definition introduces the third and last type of force discussed by Newton in this context, the centripetal force: "A centripetal force is that by which bodies are drawn or impelled or any way tend, towards a point as to a centre." [13] In Latin: "Vis centripeta est, qua corpora versus punctum aliquod tanquam ad centrum undique trahuntur, impellantur, vel utcunque tendunt." Why does Newton supply a special definition for centripetal force if this kind of force is to be considered, in accordance with the concluding remarks of Definition IV, on the same footing as percussion or pressure? Why do we not find any definition

[12] *Principles,* p. 2.
[13] *Principles,* p. 2.

of the force of percussion or the force of pressure? It seems that Newton regarded centripetal force as of greater importance than all the other forces; it occupied his mind more than anything else in mechanics and it was his real point of departure for all considerations of this chapter of definitions. Indeed, his remarks on this definition begin with the words: "Of this sort is gravity."

The next three definitions define the absolute, accelerative and motive quantities of centripetal forces:

Definition VI: The absolute quantity of a centripetal force is the measure of the same, proportional to the efficacy of the cause that propagates it from the centre through the spaces round about.

Definition VII: The accelerative quantity of a centripetal force is the measure of the same, proportional to the velocity which it generates in a given time.

Definition VIII: The motive quantity of a centripetal force is the measure of the same, proportional to the motion which it generates in a given time.

These quantities are called by Newton, for the sake of brevity, absolute, accelerative, and motive forces. In order to understand his interpretation of these forces most clearly, let us consider the electrostatic forces exerted on the surrounding medium by a charged body at the center. The charge itself corresponds roughly to what Newton calls "absolute force." "I refer," he says, "the absolute force to the centre, as endued with some cause, without which those motive forces would not be propagated through the spaces round about; whether that cause be some central body (such as is the magnet in the centre of the magnetic force, or the earth in the centre of the gravitating force), or anything else that does not yet appear."[14] Accelerating force, in Newton's conception, may then be taken as corresponding to the force, in the conventional modern meaning, exerted on a unit of mass, that is, numerically equal to the acceleration. Newton refers "the accelerative force to the place of the body, as a certain power diffused from the centre to all places around to move the bodies

[14] *Ibid.*, p. 5.

that are in them,"[15] words which seem to suggest that Newton already was thinking of force in the conception of field. Finally, motive force would correspond to the force — again in the conventional meaning as mass times acceleration — exerted on the charged test body in the electrostatic field. In fact, Newton says: "Wherefore the accelerative force will stand in the same relation to the motive, as celerity does to motion."[16] Needless to say, little use is made of the concept of absolute force, since its intensity is ascertainable only by accelerative or motive forces. Indeed, Newton seems to have already discarded this notion altogether in his own work.

For further reference, and for the sake of completeness, we now cite Newton's three famous axioms of motion without going into a detailed critical discussion of them here, since that forms part of the nineteenth-century criticism of the Newtonian concept of force.

Law I: Every body continues in its state of rest, or of uniform motion in a right line, unless it is compelled to change that state by force impressed upon it.

Law II: The change of motion is proportional to the motive force impressed; and is made in the direction of the right line in which that force is impressed.

Law III: To every action there is always opposed an equal reaction; or, the mutual actions of two bodies upon each other are always equal, and directed to contrary parts.[17]

The first two laws of motion, which Newton credits to Galileo and Huygens, add little information on Newton's conception of force.

Because of their fundamental historical importance Newton's original text will be quoted here:

Lex I (editions of 1687 and 1713): Corpus omne perseverare in statu suo quiescendi vel movendi uniformiter in directum, nisi quantenus a viribus impressis cogitur statum illum mutare.

[15] *Ibid.*

[16] *Ibid.*

[17] *Principles,* p. 13.

Lex I (edition of 1726): Corpus omne perseverare in statu suo quiescendi vel movendi uniformiter in directum, nisi quatenus illud a viribus impressis cogitur statum suum mutare.

Lex II: Mutationem motus proportionalem esse vi motrici impressae, et fieri secundum lineam rectam qua vis illa imprimitur.

Lex III: Actioni contrariam semper et aequalem esse reactionem: sive corporum duorum actiones in se mutuo semper esse aequales et in partes contrarias dirigi.

The first axiom, the principle of inertia, may either be interpreted as a qualitative definition of force or, if force is taken as recognisable in a manner independent of the laws of motion, as an empirical statement describing the motion of free bodies. The second law, likewise, has two possible interpretations: it may serve as a quantative definition of force or as a generalization of empirical facts.[18] In modern notation the law, according to Newton, asserts $F \propto \Delta(mv)$.

Since Newton clearly distinguishes between definitions and axioms (or laws of motion), it is obvious that the second law of motion was not intended by Newton as a definition of force, although it is sometimes interpreted as such by modern writers on the foundations of mechanics. Nor was it meant to be merely the statement of a method of measuring forces. Force, for Newton, was a concept given a priori, intuitively, and ultimately in analogy to human muscular force. Definition IV, therefore, is not to be interpreted as a nominal definition, but as summarizing the characteristic property of forces to determine accelerations.

Yet, at the end of the discussion following Definition VIII, Newton declares:

I refer the motive force to the body as an endeavor and propensity of the whole towards a centre, arising from the propensities of the several parts taken together . . . and the absolute force to the centre, as endued with some cause, without which those motive forces would not be propagated through the spaces round about; whether that cause be some

[18] Cf. Mary B. Hesse, *Forces and Fields* Thomas Nelson, London, 1961 pp. 134 ff.

central body (such as is the magnet in the centre of the magnetic force, or the earth in the centre of the gravitating force), or anything else that does not yet appear. For I here design only to give a mathematical notion of those forces without considering their physical causes and seats.[19]

A few lines later Newton stresses the point again: "considering those forces not physically, but mathematically." These remarks seem to indicate that Newton was already groping after a somewhat more positivistic interpretation of the concept of force.

Newton's ultimately undecided position with respect to the real nature of force in general, and also, as we shall see later on, with respect to gravitational forces in particular, shows the incredible greatness of his genius. Not only did he establish the first self-contained system of physical causality, capable of representing the deeper features of physical experience and, in addition, highly satisfactory from the logical point of view, but he surpassed himself, so to speak, in conceiving the problematic character of the fundamental notions of his system. Einstein, on the occasion of the two hundredth anniversary of Newton's death, declared: "Newton himself was better aware of the weaknesses inherent in his intellectual edifice than the generations of learned scientists which followed him. This fact has always aroused my deep admiration."[20]

Newton's expression of force as rate of change of momentum seems to suggest that he was led to this formulation by the study of the laws of impact of solid bodies. As a matter of fact, the phenomena of impact and collision were the subject of most intensive study during the decades preceding the publication of the *Principia;* Galileo had already recognized the importance of experiments on percussion for a clear comprehension of the concept of force and wrote a special treatise on this subject,

[19] *Principles,* p. 5.

[20] Albert Einstein, "The mechanics of Newton and their influence on the development of theoretical physics," reprinted and translated from an article in *Die Naturwissenschaften,* vol. 15 (1927) in *Ideas and opinions* by Albert Einstein (Crown Publishers, New York, 1954), p. 257.

entitled "On the force of percussion." [21] It is here that Galileo for the first time brings force in relation with momentum (we ignore his confounding of mass and weight): "It is manifest," he says, "that the force of the mover or the resisting object is not a simple concept, but composed of two ideas which together determine the energy to be measured; one is the weight [mass], the other speed." [22] It has even been suggested that Descartes's identification of force with momentum might have had its origin in this passage, a conjecture that is unlikely, since the Elzevir edition, which alone could have been accessible to Descartes, did not contain this treatise. We also know that as early as 1639, when Marcus Marci, the rector of the University of Prague, published his *De proportione motu*[23] dealing with the problem of impact, the interest of the great theoreticians of mechanics was focused on these problems; and, in 1668, encouraged by a request of the Royal Society, Wallis, Wren, and Huygens found the final solution of the problem of impact, Wallis for inelastic bodies, Wren and Huygens for elastic bodies. No doubt, Newton had read their communications to the Royal Society and it may well be that under their influence Newton formulated his second law by means of the concept of momentum, although he undoubtedly recognized the fundamental importance of the Galilean concept of acceleration in this respect. The *ultima ratio* in his reasoning was, of course, the law of inertia. To make it quite clear, we do

[21] Galileo Galilei, "Della forza della percossa," sometimes added as the sixth day (*giornata sesta*) to his *Dialogues concerning two new sciences*. Cf. the German translation in Ostwald's *Klassiker* (Leipzig, 1890–1891), No. 25. The original Leiden edition did not contain Galileo's discussion on percussion, primarily because the Elzevirs hastened to publish the work at the earliest possible date.

[22] "È manifesto, la facultà della forza del movente e della resistenza del mosso non essere una e semplice, ma composta di due azioni, dalle quali la loro energia dee essere misurata; l'una delle quali è il peso, si del movente come del resistente, e l'altra è la velocità, secondo la quale quello dee muoversi e questo esser mosso." *Le opere de Galileo Galilei* (national edition; Florence), vol. 3 (1892), p. 329.

[23] The laws of impact of elastic bodies are derivable from Newton's laws of motion, but not vice versa.

not suppose that Newton inferred his second law from the laws of impact; such an inference would have met with insuperable conceptual and mathematical difficulties. Ultimately, it was a stroke of genius, a "free creation of the human mind."

The first two laws of motion add little information on Newton's concept of force that is not contained in the preceding definition. The third law, however, supplies an additional important characteristic of force not mentioned previously: force manifests itself invariably in a dual aspect; it is action and reaction simultaneously. Much as a business transaction can be regarded both as a purchase and as a sale of the same amount, force can be considered as action as well as reaction of the same magnitude. As far as attraction is concerned, Newton believes himself capable of demonstrating the validity of the third law as follows: Suppose the two bodies *A* and *B* attract each other; imagine, now, that *A*, for example, attracts *B* with greater intensity than *B* attracts *A*; suppose also that an obstacle is interposed to prevent the meeting of these two bodies; our assumption would then lead to the conclusion that the whole system (*A*-obstacle-*B*) would move in the direction from *B* to *A*, for the obstacle, as Newton puts it, "will be more strongly urged by the pressure"[24] of the body *B* than by the pressure of the body *A* and consequently will not remain in equilibrium, but accelerate *in infinitum*.

It is generally contended that Newton's statement of the third law is "his most important achievement with respect to the principles"[25] of mechanics. Without diminishing Newton's outstanding merits in the foundation of classical mechanics in the least, it should be recalled that Kepler had already a clear conception of the reciprocity of force,[26] although he never formulated this idea in a general quantitative principle and did not realize the equality of the two forces involved and their opposite directions.

[24] *Principles*, p. 25.

[25] Ernst Mach, *The science of mechanics*, trans. T. J. McCormack (Open Court, La Salle, Ill., 1942), p. 243.

[26] At least as far as *cognata corpora* are concerned. See p. 91.

Newton's Corollary I which follows immediately after the third law of motion contains the theorem of the parallelogram of forces:

> A body, acted on by two forces simultaneously, will describe the diagonal of a parallelogram in the same time as it would describe the sides by those forces separately.[27]

Newton's formulation of the parallelogram theorem is of great importance for our understanding of his conception of force, not only because it characterizes force as a vectorial quantity, to use a modern expression, but also because it throws some light on how he conceived the precise mechanism of dynamic action at the time when he composed the introductory chapter to his *Principia.*

It is well known that Daniel Bernoulli, in a paper, "Examen principiorum mechanicae et demonstrationes geometricae de compositione et resolutione virium," published in 1726, attempted to show that the parallelogram theorem was an a priori truth, independent of observational experience. The problem of its necessary validity occupied the minds of great theoreticians of mechanics in the eighteenth century, such as d'Alembert, and even interested Siméon Denis Poisson as late as 1833, although by 1736 Leonhard Euler had made it clear, in his *Mechanica,*[28] that the theorem cannot be proved analytically without further assumptions. Newton did not yet face this problem in this extreme version, so characteristic of eighteenth-century thought; nevertheless, he was convinced that its validity can be derived from the very concept of force, as is shown by his demonstration following Corollary I. Newton's derivation of the theorem is based on the kinematic composition of velocities, known already to Aristotle,[29] and employed first intuitively by Leonardo da

[27] *Principles,* p. 14.

[28] Leonhard Euler, *Mechanica sive motus scientiae analytice exposita scientiarum* (St. Petersburg, 1736), book I.

[29] Cf. Pierre Duhem, *Les origines de la statique* (Paris, 1905–1906), vol. 2, p. 245.

Vinci, then consciously by Benedetti[30] and by Galileo.[31] In statics the principle had been employed first by Simon Stevin, who in his *De Beghinselen de Weeghconst*[32] tried to prove that three forces which in magnitude and direction correspond to the three sides of a triangle balance one another, a statement that is equivalent to the parallelogram theorem; later Giles Personne de Roberval brought it into relation with the principle of virtual work.[33]

Curiously enough, in 1687, the year of the publication of the *Principia,* two other expositions on the composition of forces were published. Pierre Varignon enounced the theorem of the parallelogram of forces in his *Projet de nouvelle mécanique*[34] and gave a more detailed discussion in his *Nouvelle mécanique ou statique.*[35] In a letter addressed to M. J. Dieulamant, a royal engineer in Grenoble, Pierre Lamy, independently of Newton and Varignon, propounds the theorem under the title "Nouvelle manière de démontrer les principaux théorèmes des éléments de mécanique." [36] Newton's French contemporaries, in their discussion on the composition of forces, conceived forces in the traditional Peripatetic manner as related to the velocity of the body on which they act, and consequently took the kinematic principle as their point of departure.

[30] Giovanni Battista Benedetti (Benedicti), *Diversarum speculationum mathematicarum et physicarum liber* (Turin, 1585), p. 160.

[31] Galileo Galilei, *Dialogues concerning two new sciences,* trans. Henry Crew and Alfonso de Salvio (Northwestern University Press, Evanston, 1946), pp. 234, 246, 256.

[32] Simon Stevin, *De Beghinselen de Weeghconst* (Leiden, 1586).

[33] Giles Personne de Roberval, "Observations sur la composition des mouvements" (1693), *Mémoires de l'Académie des Sciences, Paris, 6,* 1–89 (1730). Cf. also his essay, "Project d'un livre de mécanique traitant des mouvements composés," *ibid.,* pp. 68–71. Cf. also his "Traité de mécanique des poids," published in Mersenne's *Traité de l'harmonie.*

[34] Pierre Varignon, *Projet de nouvelle mécanique* (Paris, 1687).

[35] Pierre Varignon, *Nouvelle mécanique ou statique* (Paris, 1725), posthumously published.

[36] This letter forms part of the later editions of Lamy's *Traité de mécanique, de l'équilibre des solides et des liqueurs,* the first edition of which was published in 1679.

These coincidences show clearly that the way to add forces was well understood, however they might be defined. In other words, any definition of force must be in accordance with the parallelogram theorem. Newton, therefore, felt obliged to show that the theorem is indeed consistent with his concept of force; and the surest way to show consistency is to deduce one from the other.

In order to do so Newton begins his proof with the following words:

> If a body in a given time, by the force *M* impressed apart in the place *A* [Fig. 4], should with a uniform motion be carried from *A* to *B*, and by the force *N* impressed apart in the same place should be carried from *A* to *C*, let the parallelogram *ABCD* be completed, and, by both forces acting together, it will in the same time be carried in the diagonal from *A* to *D*.[37]

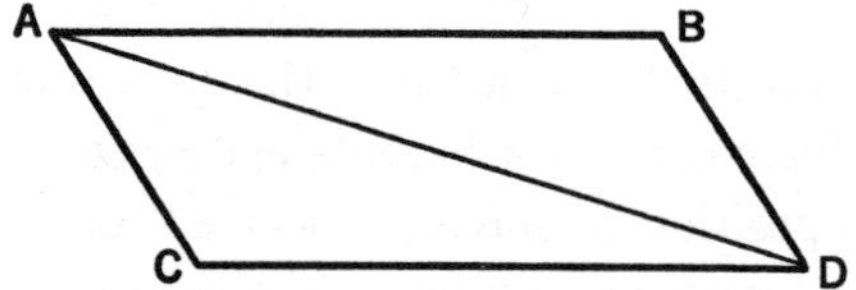

Fig. 4. The parallelogram of forces.

Newton speaks here explicitly of a uniform motion as resulting from a given force, whereas it is the change of motion, that is, acceleration times mass, that, according to the second law of motion, is proportional to the impressed force. These two apparently contradictory ideas could have been reconciled in Newton's mind only under one condition: Newton conceived the accelerative action of force as a series of successive actions that imparts to the moving object successive increments of velocity. In fact, this interpretation seems not to be preposterous if we remember his comments on his second law of motion: "If any force generates a motion, a double force will generate double the

[37] *Principles,* p. 14. In the Latin original: "Si corpus dato tempore, vi solâ *M* in loco *A* impressâ, ferretur uniformi cum motu ab *A* ad *B* . . ."

motion, a triple force triple the motion, whether that force be impressed altogether and at once, or gradually and successively."[38] Ultimately it is the old Cartesian conception of force which, intentionally or unintentionally, creeps in again and is employed for the derivation of the principle of composition of forces.

As a matter of fact, the conception of force as a cumulative effect has a history of its own. Leonardo da Vinci seemed to have embraced this view already when he remarked occasionally that the arrow, released from the bow, is not projected only at the moment of the greatest tension, but also receives additional impulses at the other positions of the strained string.[39] The force impressed on the flying arrow is thus but a sequence of consecutive impulses, similar to the action of gravity, which in the middle of the sixteenth century and later, by Alessandro Piccolomini, Julius Caesar Scaliger, and Galileo's predecessor Giovanni Battista Benedetti, was conceived as communicating consecutive impulses to a falling body. It also seems most probable that Descartes when studying together with Beeckmann the phenomena of free fall, during his residence in Holland, was influenced by Scaliger and Benedetti and consequently adopted this concept of force, as far as one can speak of a Cartesian concept of force. Still, in this early analysis of free fall he speaks of an attractive force exerted by the earth on the falling stone, "Quod ut demonstrem, assumam pro primo minimo vel puncto motus, quod causatur a prima quae imaginari potest attractiva vi terrae."[40] In his physico-mathematical notes, from which these passages are quoted, he describes the operation of this force as created by God in a discrete sequence of impulses: "Deus creet . . . vim attractivam lapidis; et singulis postea momentis novam et novam vim creet, quae aequalis sit illi quam primo

[38] *Ibid.*, p. 13.

[39] Cf. Pierre Duhem, "De l'acceleration produite par une force constante; notes pour servir a l'histoire de la dynamique," *Congrès international de philosophie* (Geneva, 1905), p. 882.

[40] *Oeuvres de Descartes,* vol. 10 (1908), p. 75.

momento creavit; quae juncta cum vi ante creata fortius lapidem trahat."[41] It is interesting to note that this notion of force as a series of impacts not only affected Newton's concept of force, as we have seen, but also had its repercussions on John Bernoulli's concept of force early in the eighteenth century, and even on Lazare Carnot's[42] toward the end of that century.

Newton's derivation of the parallelogram theorem of forces tacitly assumes that the action of one force on a body is independent of the action of another force, an assumption that is far from self-evident. In fact, this very supposition has been called in question by various theoreticians when treating the problem of molecular forces. Thus in particular Barré de Saint-Venant, a convinced atomist in the middle of the last century, who subjected the general Newtonian conception of force to severe criticism in his *Principes de mécanique fondés sur la cinématique*,[43] seeking a theoretical basis for the theory of elasticity and a molecular theory for the study of the strength of materials, did not hesitate to reject the validity of the parallelogram theorem for the microscopic realm. In one of his last publications, "De la constitution des atomes," he stated emphatically:

[41] *Ibid.*, pp. 77–78.

[42] "La pesanteur et toutes les forces de ce genre opèrent par degrés insensibles et ne produisent aucun changement brusque. Cependant il paraît assez naturel de les considérer comme imprimant, à des intervalles infiniment petits, des coups eux-mêmes infiniment petits aux mobiles qu'elles animent." Lazare Carnot, *Essai sur les machines en général* (Paris, 1783); *Principes généraux de l'équilibre et du mouvement* (Paris, 1803), p. 2.

[43] Barré de Saint-Venant, *Principes de mécanique fondés sur la cinématique* (Paris, 1851). John Bernoulli, *Opera omnia,* ed. G. Cramer (Lausanne and Geneva, 1742), vol. 4, p. 256, stated: "Peccant qui compositionem virium cum compositione motuum confundunt." In 1852 Ferdinand Reech, *Cours de mécanique* (Paris, 1852), p. 61, declared: "Nous rejetterons absolument toutes les prétendues démonstrations du théorème du parallélogramme de forces au moyen de la règle évidente du parallélogramme des vitesses en géometrie." The logical status of the parallelogram theorem was thoroughly investigated by Gaston Darboux, *Bull. scienc. math. 9,* 281 (1875). For further details the reader is referred to Ernst Goerges, *Die Zusammensetzung der Kräfte* (Halle, 1909) and to Schimmak's *Axiomatische Untersuchungen über die Vektoraddition* (Halle, 1908).

On remarquera qu'elle entraîne aussi que la force totale sollicitant une particule n'es pas exactement la résultante géométrique, composée par la règle statique du parallélogramme ou du polygone que l'on connait, de toutes les forces avec lesquelles la solliciteraient séparément les autres particules si chacune existait seul avec elle, comme on l'a cru jusqu'à nos jours; cette règle ne serait plus vraie que pour les actions à des distances perceptibles, dont l'intensité, réciproque aux carrés des distances, est celle de la pesanteur universelle, toujours négligeable vis-à-vis des actions à des distances imperceptibles qui produisent l'élasticité, la capillarité, les chocs, les pressions et les vibrations.[44]

These ideas were not confined, moreover, to French elasticians, as can be seen from John Jellet's theory of "modified action."[45] That Newton's homeland had its share in this criticism of the parallelogram theorem is stressed also by Lord Kelvin and Peter Guthrie Tait in their elaborate *Treatise on natural philosophy:*

Though Newton perceived that the Parallelogram of Forces, or the fundamental principle of Statics, is essentially involved in the second law of motion, and gave a proof which is virtually the same as the preceding, subsequent writers on Statics (especially in this country) have very generally ignored the fact; and the consequence has been the introduction of various unnecessary Dynamical Axioms, more or less obvious, but in reality included in or dependent upon Newton's laws of motion.[46]

Turning now to Newton's concept of the force of gravitation, the force par excellence, we shall discuss the historical and biographical sidelights associated with this most famous of Newton's discoveries only so far as they have a bearing on Newton's conception of universal attraction. The subject is highly controversial in two respects. First, historians of science

[44] Barré de Saint-Venant, "De la constitution des atomes," *Annales de la Société scientifique de Bruxelles 2* (1878). Cf. Isaac Todhunter, *A history of the theory of elasticity,* ed. Karl Pearson (Cambridge University Press, Cambridge, 1886), vol. 2, p. 183.

[45] John H. Jellet, "On the equilibrium and motion of an elastic solid," *Transactions of the Royal Irish Academy 22,* 179 (1855).

[46] Lord Kelvin and P. G. Tait, *Treatise on natural philosophy* (Cambridge University Press, Cambridge, reprinted 1923), part I, p. 244. Kelvin and Tait reformulated Newton's second law to include essentially the parallelogram theorem.

are divided in their opinions as to whether Newton conceived gravitation as a pure action at a distance or not. Second, in recent years much has been published on a possible direct influence upon Newton by Jacob Böhme, the greatest of Protestant mystic writers of the seventeenth century. In his conception of universal attraction, it is claimed, "Newton was directly influenced by Jacob Böhme's mysticism." [47] Such an authority as Professor E. N. da Costa Andrade, to mention another reference, wrote in 1946: "Newton was a close student of Jacob Böhme, from whose works he copied large extracts." [48] If this assumption could be substantiated and vindicated beyond doubt, Newton's conception of gravitation would be traceable to the ancient Neo-Platonic and gnostic conceptions of force as *dynamis*,[49] via their mystic interpretations by Böhme.

In order to form our own judgment on these questions, let us first of all study Newton's own statements about gravitation. We do not claim our list of references to be exhaustive, but let us try to analyze the most important passages in Newton's works and correspondence as far as they are relevant for our subject.

Newton states in a letter[50] to Halley that his interest in the problem of gravitation was roused in 1665. As early as 1666 he was apparently in possession of the central ideas of his theory of gravitation. In a letter to Henry Oldenburg, written in 1675, Newton discusses some vague speculations on an ethereal medium as a possible explanation of gravity.

But to proceed to the hypothesis: It is to be supposed therein, that there is an aetherial medium, much of the same constitution with air, but far rarer, subtiler, and more strongly elastic . . . For the electric

[47] A. J. Snow, *Matter and gravity in Newton's physical philosophy* (Oxford University Press, London, 1926), p. 193.

[48] E. N. da C. Andrade, "Newton," in *The Royal Society, Newton Tercentenary Celebrations, 15–19 July 1946* (Cambridge University Press, Cambridge, 1947), p. 20.

[49] See p. 49.

[50] The letter, dated June 20, 1686, is given in full in the appendix to W. W. Rouse Ball, *An essay on Newton's* Principia (London, 1893). Cf. Brewster, *Memoirs of . . . Newton,* vol. 1, p. 295.

and magnetic effluvia, and the gravitating principle, seem to argue such variety. Perhaps the whole frame of nature may be nothing but various contextures of some certain aetherial spirits or vapours, condensed as it were by precipitation, much after the manner that vapours are condensed into water, or exhalations into grosser substances, though not so easily condensable, and after condensation wrought, into various forms, at first by the immediate hand of the Creator, and ever since by the power of nature, which, by virtue of the command increase and multiply, became a complete imitator of the copy set her by the Protoplast. Thus perhaps may all things be originated from aether.[51]

This passage, in which Newton describes a tentative unified theory of all physical phenomena, is reminiscent of the style of the Cambridge Platonists Cudworth and More, of whom we shall have to speak more in the following chapter. A more detailed description of the supposed ether and a more deliberate speculation on the cause of gravity is expounded in a letter to Robert Boyle, dated February 28, 1678. Here Newton says:

I shall set down one conjecture more . . . it is about the cause of gravity. For this end I will suppose aether to consist of parts differing from one another in subtility by indefinite degrees: that in the pores of bodies, there is less of the grosser aether in proportion to the finer, than in open spaces; and consequently, that in the great body of the earth there is much less of the grosser aether, in proportion to the finer, than in the regions of the air; and that . . . from the top of the air to the surface of the earth, and again from the surface of the earth to the centre thereof, the aether is insensibly finer and finer. Imagine, now, any body suspended in the air, or lying on the earth; and the aether being, by the hypothesis, grosser in the pores which are in the upper parts of the body, than in those which are in the lower parts; and that grosser aether, being less apt to be lodged in those pores, than the finer aether below; it will endeavour to get out, and give way to the finer aether below, which cannot be, without the bodies descending to make room above for it to go out into.[52]

Newton's attempt to deduce the phenomenon of gravity is based, chiefly, on two assumptions: a stratification of elastic ether particles of continuously varying subtility, surrounding and

[51] Brewster, *Memoirs of . . . Newton,* vol. 1, p. 392.

[52] *Isaaci Newtoni Opera,* ed. Samuel Horsley (London, 1779–1785), vol. 4, p. 385. A more complete extract of this letter is given by Florian Cajori in his explanatory appendix to the *Principles,* p. 675.

permeating the earth; and their irreducible tendency to dilatation. The first assumption is, of course, based on Cartesian cosmology which since 1644 was the dominant doctrine in scientific circles both on the continent and in England. Newton followed the line, but already at this date cautioned his reader not to take these speculations too seriously, as he says, "my notions about things of this kind are so indigested, that I am not well satisfied myself with them."

At the end of 1675 Newton composed "A theory of light and colours,"[53] which was sent to the Royal Society at the request of Oldenburg and forms the nucleus of his *Opticks,* first published in 1704. In this paper, in which optical phenomena are explained as resulting from violent motions of "ethereal spirits," ether particles are referred to as explanatory principles for numerous physical and even biological processes, such as the contraction and dilatation of muscles. Gravity, too, is interpreted by the mechanism of ethereal particles in much the same manner as in the letter to Boyle. His conception of gravity may perhaps be best understood, if, to use a modern illustration, the motion of gravitating matter may be compared to the motion of a diamagnetic specimen in a magnetic field of locally varying intensity.

That Newton did not also reject the possible existence of an ethereal medium, instrumental for an essentially kinetic theory of gravity, at the time when he composed the first edition of the *Principia* is obvious from his remarks on Definition I, where he says: "I have no regard in this place to a medium, if any such there is, that freely pervades the interstices between the parts of bodies."[54]

"If any such there is"—these words are characteristic of

[53] Isaac Newton, "A theory of light and colours, containing partly an Hypothesis to explain the properties of light discoursed by him in his former papers, partly the principal phenomena of the various colours . . ." in T. Birch, *The history of the Royal Society of London* (London, 1756–1757), vol. 3, p. 247.

[54] *Principles,* p. 1.

Newton's attitude toward the problem of the nature of gravitation as far as the publication itself is concerned. Not in possession of sufficient experimental or observational material to substantiate a specific theory on the nature of gravitation, Newton preferred to abstain from passing judgment on this question and treated gravitation independently of whether it is an action at a distance or is caused by contiguous action between ether particles and ordinary matter. While he speaks of "bodies attracting each other," [55] "the attractions of one corpuscle towards the several particles of one sphere," [56] of "mutual attraction," [57] and uses similar expressions that could easily have misled the reader to the assumption that he conceived of the forces involved as innate in matter and acting at a distance, Newton nowhere in the first edition of the *Principia* made a statement to this intent.

Furthermore, in the second edition of the *Principia* (1713) he made an explicit declaration, perhaps after having realized the possibility of causing such misconceptions, and said:

> I here use the word *attraction* in general for any endeavor whatever, made by bodies to approach each other, whether that endeavor arise from the action of the bodies themselves, as tending to each other or agitating each other by spirits emitted; or whether it arises from the action of the ether or of the air, or of any medium whatever, whether corporeal or incorporeal, in any manner impelling bodies placed therein towards each other.[58]

In the beginning of Book III ("System of the world") in the chapter discussing the rules of reasoning in philosophy, Newton added in the second edition: "Not that I affirm gravity to be essential to bodies: by their *vis insita* I mean nothing but their inertia." [59] The sudden resumption of the concept of *vis insita* and its renewed interpretation as inertia was called forth, in Newton's opinion, to neutralize his former expressions, that is,

[55] *Principia,* Book I, Proposition LX; *Principles,* p. 167.

[56] *Principia,* Book I, Proposition LXXII; *Principles,* p. 195.

[57] *Principia,* Book III, Proposition V; *Principles,* p. 410.

[58] *Principia* (second edition), Scholium following Proposition LXIX, Book I; *Principles,* p. 192.

[59] *Principles,* p. 400.

Definition III, in which inertia was called an "innate force of matter," "a power of resisting,"[60] and his remarks on inertia in Rule III, where he stated that all bodies are "endowed with certain powers (which we call the inertia)."[61] Finally, we have Newton's famous "hypotheses non fingo" at the end of Book III in the General Scholium, where he says emphatically:

> Hitherto we have explained the phenomena of the heavens and of our sea by the power of gravity, but have not yet assigned the cause of this power . . . But hitherto I have not been able to discover the cause of those properties of gravity from phenomena, and I frame no hypotheses; for whatever is not deduced from the phenomena is to be called an hypothesis; and hypotheses, whether metaphysical or physical, whether of occult qualities or mechanical, have no place in experimental philosophy. In this philosophy particular propositions are inferred from the phenomena, and afterwards rendered general by induction. Thus it was that the impenetrability, the mobility, and the impulsive force of bodies, and the laws of motion and of gravitation, were discovered. And to us it is enough that gravity does really exist, and act according to the laws which we have explained, and abundantly serves to account for all the motions of the celestial bodies, and of our sea.[62]

At the same time, however, Newton alludes at the end of this famous General Scholium to the idea of "ethereal spirits," when he says:

> And now we might add something concerning a certain most subtle spirit which pervades and lies hid in all gross bodies; by the force and action of which spirit the particles of bodies attract one another at near distances, and cohere, if contiguous . . . But these are things that cannot be explained in few words, nor are we furnished with that sufficiency of experiments which is required to an accurate determination and demonstration of the laws by which this electric and elastic spirit operates.[63]

Newton's "ethereal spirits" were misinterpreted as "immaterial spirits or principles" and were eagerly employed already by his contemporaries for a rational foundation of their theistic doctrines. Most famous among these was Bentley, who stretched the

[60] *Ibid.*, p. 2.
[61] *Ibid.*, p. 399.
[62] *Ibid.*, p. 546.
[63] *Ibid.*, p. 547.

meaning of Newton's conception of gravitation in this sense most eloquently in his *Sermons*. Mutual attraction without contact or impulse, he contended, is not an attribute of mere matter but must be due to an immaterial principle, an immaterial living mind that must "inform and actuate the dead matter, and support the frame of the world." Universal gravitation for Bentley was above all mechanism or material causes, and "proceeds from a higher principle, a divine energy and impression."[64]

Bentley's interpretation of universal gravitation did not meet with Newton's approval. In a letter to Bentley, dated 1692, Newton opposes Bentley's assertions with the following words:

> You some times speak of gravity as essential and inherent to matter. Pray, do not ascribe that notion to me; for the cause of gravity is what I do not pretend to know, and therefore would take more time to consider of it.[65]

Similarly, in a second letter Newton informs Bentley:

> It is inconceivable, that inanimate brute matter, should, without the mediation of something else, which is not material, operate upon and affect other matter without mutual contact, as it must be, if gravitation, in the sense of Epicurus, be essential and inherent in it. And this is one reason why I desired you would not ascribe innate gravity to me. That gravity should be innate, inherent, and essential to matter, so that one body may act upon another at a distance through a vacuum, without the mediation of any thing else, by and through which their action and force may be conveyed from one to another, is to me so great an absurdity, that I believe no man, who has in philosophical matters a competent faculty of thinking, can ever fall into it. Gravity must be caused by an agent acting constantly according to certain laws; but whether this agent be material or immaterial, I have left to the consideration of my readers.[66]

It is important for what follows to note that these letters to Bentley were published many years after the publication of the *Principia* and could therefore not have had any immediate influence on Newton's commentators in the eighteenth century.

[64] Richard Bentley, "Sermons preached at Boyle's lecture," in Alexander Dyce, ed., *The works of Richard Bentley* (London, 1838), vol. 3, pp. 149, 165.
[65] *Ibid.*, p. 210.
[66] *Ibid.*, Letter III, p. 211.

Furthermore, his remarks in the General Scholium, in which he gives free reign to his religious enthusiasm, in which he speaks of God as a Being in whom "are all things contained and moved," [67] gave additional material to his theistic interpreters for substantiating their doctrines on Newton's theory of gravitation.

We now come to the final document in this list of references of Newton's own statements about gravitation, to the *Opticks*, published in its first edition in 1704. Of greater importance for our discussion, however, is the second edition (1717), in the Advertisement to which Newton says:

> And to shew that I do not take Gravity for an essential Property of Bodies, I have added one Question concerning its Cause, chusing to propose it by way of a Question, because I am not yet satisfied about it for want of Experiments.[68]

Thus, in Question 31 Newton states again:

> How these Attractions may be perform'd, I do not here consider. What I call Attraction may be perform'd by impulse, or by some other means unknown to me. I use that Word here to signify only in general any Force by which Bodies tend towards one another, whatsoever be the Cause.[69]

Here, again, we have the somewhat agnostic attitude found in the *Principia* with respect to the cause of gravitation. It must, however, be admitted that in the *Opticks*, too, certain statements seem to suggest that Newton, at times at least, was not so strongly opposed to the conception of gravitation as an innate power in matter. Thus in Question 31 he writes:

> All these things being consider'd, it seems probable to me, that God in the Beginning form'd Matter in solid, massy, hard, impenetrable, moveable Particles, of such Sizes and Figures, *and with such other Properties*,[70] and in such Proportion to Space, as most conduced to the End for which he form'd them . . . It seems to me farther, that these

[67] *Principles*, p. 545.

[68] Sir Isaac Newton, *Opticks, or a treatise of the reflections, refractions, inflections and colours of light* (reprinted from the fourth edition, London, 1931), Advertisement II.

[69] *Ibid*., p. 376.

[70] Italics mine.

Particles have not only a *Vis inertiae,* accompanied with such passive Laws of Motion as naturally result from that Force, but also that they are moved by certain active Principles, such as is that of Gravity.[71]

It may not be impossible that, when he inclined in the latter part of his life more definitely to a corpuscular theory of light and thought the assumption of an all-pervading ether superfluous, this change of attitude had its repercussion also upon his conception of gravitation. On the whole, however, as far as scientific methodology was concerned he was opposed to any metaphysical or theological interpretation of gravitation, in sharp contrast to his attitude toward the conception of space, an attitude that, as I have shown in *Concepts of space,*[72] was strongly interlocked with metaphysical and theological considerations.

To sum up: the concept of gravitational force is an ultimately irreducible notion in Newton's conceptual scheme of physical science. It is distinguished from other kinds of force by its universality and consequent importance for astronomical and cosmological considerations. Its quantitative aspects are ascertained from experimental observation; its ultimate nature is unknown.

It is important to note that this implied for Newton's contemporaries that gravitation remains an unexplained phenomenon. For to explain a physical phenomenon, at that time, still meant to give its mechanical[73] causes, that is, to account for it in terms of the fundamental qualities of matter, impenetrability, extension, inertia, together with the laws of motion. Now, it was clear that the traditional mechanical explanation, as advanced

[71] Newton, *Opticks,* pp. 400–401.

[72] Max Jammer, *Concepts of space, the history of theories of space in physics* (Harvard University Press, Cambridge, 1954), pp. 93–124.

[73] It is not easy to find in the literature of that period an exact definition of the term "mechanical." Roger Cotes, in his preface to the *Principia* (*Principles,* p. xxvii), states that a quality is said to be mechanically caused when it is produced by some of the other affections of body. In general, "mechanical" meant "in accordance with the laws of motion." See, for example, George Cheyne, *Philosophical principles of religion, natural and revealed* (London, 1715), p. 24. Cf. also Colin MacLaurin, *An account of Sir Isaac Newton's philosophical discoveries* (London, 1748), p. 241.

by Descartes or by Huygens, could not account for the fact that gravitation is proportional to the masses and not to the surfaces of the bodies concerned. In fact, Newton recognized[74] that all theories based on mere external impact were basically insufficient because of that reason. And yet it was the only available scientific method of explanation that was at the time unimpeachable from the methodological point of view. These considerations, most certainly, are responsible for Newton's indecision and doubt whether gravitation could ever be reduced to purely mechanical phenomena.

Having thus clarified Newton's conception of gravitation, let us now turn to the historical problem of a possible direct influence of Jacob Böhme upon this conception. First of all, we consider it wholly unfounded that Böhme inspired Newton in his axiomatization of mechanics, that is, in the composition of the laws of motion, and reject Spurgeon's contention that "it is almost certain that the idea of the three laws of motion first reached Newton through his eager study of Böhme." [75] Consequently, we confine our discussion to Newton's conception of universal attraction.

It would lead us too far afield and would increase the length of the present chapter beyond all reasonable limits should we give a detailed account of Böhme's obscure mystical system of the seven spirits or *Quellengeister*[76] on which his analysis of the whole of manifested existence is founded. The first spirit or quality among these, according to Böhme, is the universal property of all that which compresses, astringes, and hardens; it is the principle of inward attraction or contraction, a primordial desire to coagulate itself into matter. The second quality or

[74] The force of gravitation acts "not according to the quantity of the surfaces of the particles upon which it acts (as mechanical causes use to do)," *Principles*, p. 506.

[75] Caroline F. E. Spurgeon, "William Law and the mystics," in *The Cambridge history of English literature* (Cambridge, 1912), vol. 9, p. 307.

[76] *Quelle* (source, fount) is in this context a mystical alliteration to the concept of "quality."

spirit is action in the opposite direction, a tendency to expansion. For further information we refer the reader to the writings of the Silesian mystic, in particular to his *Threefold life, Mysterium magnum,* and *Signatura rerum.* Böhme here conceives nature as a harmonious equilibrium between these seven universal forces and makes frequent use of mechanical and physical analogies, a fact that is comprehensible if we consider his intellectual milieu which associated theological and psychological conceptions inextricably with astrological and alchemistic ideologies.

It is true that Böhme's allegorico-naturalistic writings had been translated and published in England during the seventeen years following the first English edition of his *Two theosophicall epistles*[77] in 1645, and Charles Hotham, John Sparrow, and later even Henry More, contributed much to increase the knowledge of Böhme in that country. According to Popp[78] and to Adams, a librarian of Trinity College, we know that an anthology of Böhme's writings, the *Mercurius Teutonicus,*[79] was in that library in Newton's time. Yet none of Böhme's writings were found in Newton's possession at the time of his death.[80] In fact, there is no convincing evidence that Newton ever read even one single work of Böhme's.

The tradition that identifies Jacob Böhme as the intellectual source of Newton's conception of force can be traced back to William Law, a fellow of Emmanuel College, Cambridge, who in his "Spirit of love" wrote with reference to Böhme's doctrine of the seven spirits:

> Here also, that is, in these properties of the desire, you see the ground and reason of the three great laws of matter and motion, lately dis-

[77] Jacob Böhme, *Two theosophicall epistles: wherein the life of a true Christian is described . . . whereunto is added a dialogue between an enlightened and a distressed Soule* (London, 1645).

[78] Karl Robert Popp, *Jakob Böhme und Isaac Newton* (inaugural dissertation; Gerhardt, Leipzig, 1935).

[79] *Mercurius Teutonicus: or Christian Information concerning the last Times . . . gathered out of mysticall writings of . . . Jacob Behmen* (London, 1649).

[80] H. MacLachlan, *Sir Isaac Newton: Theological Manuscripts* (Liverpool University Press, Liverpool, 1950), p. 20.

covered, and so much celebrated; and need no more to be told, that the illustrious Sir Isaac ploughed with Behmen's heifer, when he brought forth the discovery of them. In the mathematical system of this great philosopher, these three properties, attraction, equal resistance, and the orbicular motion of the planets as the affects of them, etc., are only treated of as facts and appearances, whose ground is not pretended to be known. But in our Behmen, the illuminated instrument of God, their birth and power in eternity is opened; their eternal beginning is shown.[81]

Law's contention of Newton's indebtedness to Böhme seems to be based on an oral statement which informed Law that at the time of Newton's death there were found "among his papers large abstracts out of J. Behmen's works written in his own hand." But, as mentioned before, no further evidence is available of the existence of such extracts and, needless to say, they have never been recovered or even mentioned by anyone else. Yet, Law claims:

This I have from undoubted authority; as also that, in the former part of his life, he (Newton) was led into a search for the Philosopher's Tincture from the same author. My vouchers are names well known and of great esteem with you. It is evidently plain that all that Sir Isaac has said of the universality, nature and effects of attraction and of the three first laws of nature, was not only said but proved in its true and deepest ground by Jacob Behmen in his Three First Properties of Eternal Nature, and from thence they are derived into his Temporal Outbirth.[82]

This is the whole documentary evidence for a long tradition of Böhme's direct influence upon Newton. It served Richard Symes,[83] C. W. Heckethorn,[84] Friedrich Christoph Oetinger,[85]

[81] William Law, *The spirit of love,* in *Works* (London, 1762), vol. 8, p. 38.

[82] The above quotation is part of Law's letter to Dr. George Cheyne, Scottish physician and mystic. The letter is quoted in full in Christopher Walton, *Notes and materials for an adequate biography of . . . W. Law* (privately printed, London, 1854), p. 46. Cf. also *The Gentleman's Magazine,* July 1782, p. 329 and December 1782, p. 575.

[83] Richard Symes, *Fire analysed, or the several parts of which it is compounded* (Bristol, 1771): Newton "certainly borrowed his attractive and repulsive powers from Behmen."

[84] Charles William Heckethorn, *Athenaeum* (1867), p. 128.

[85] *Des Wirttembergischen Prälaten Friedrich Christoph Oetinger, Sämtliche Schriften,* ed. K. C. E. Ehmann (Stuttgart, 1858), part II, pp. 198, 209.

and many others[86] as the source for their assertion of Newton's intellectual dependence on Böhme.

Even Louis T. More, the distinguished biographer of Newton, thought this tradition an established fact,[87] but has altered his view in this respect recently, as reported by Hobhouse.[88] Finally, Edwin Arthur Burtt refers to Böhme as having been "read copiously" by Newton and having strengthened Newton's "conviction that the universe as a whole is not mechanically but only religiously explicable." [89]

In conclusion it should be noted that Hobhouse,[90] who investigated this tradition thoroughly, has recently disproved its veracity altogether. But even if Newton had read Böhme's writings, we think, it would hardly be possible, after all that has been quoted, to speak of an intellectual dependence, as far as the Newtonian conceptual scheme is concerned. Böhme's pantheistic-dualistic thought is fundamentally different from Newton's theological outlook, not to mention his almost positivistic approach in mathematical physics.

For the subsequent development of physics, Newton's somewhat positivistic, or at least critical, attitude with respect to the interpretation of gravitation as action at a distance had actually little influence. In spite of early opposition, as voiced particularly by Leibniz, the notion of force as action at a distance became a basic concept for the great classical edifice of theoretical mechanics. Thus Laplace, in his *Mécanique céleste,* stated that the objective of his study is a reduction of all mechanical phenomena to forces acting at a distance. Lagrange's monumental work, *Mécanique analytique,* the highlight of classical mechanics, was written in the same spirit. The mechanics of action at a distance

[86] For further information see Popp, reference 78.

[87] L. T. More, *Isaac Newton* (New York, 1934), p. 158.

[88] William Law, *Selected mystical writings,* rev. Stephen Hobhouse (Harper, New York, 1948), p. 422.

[89] E. A. Burtt, *The metaphysical foundations of modern physical science, a historical and critical essay* (Doubleday, New York, 1954), p. 202.

[90] Reference 88.

gained further support in the successful applications of the classical theories of electricity and magnetism, as expounded by Laplace and Poisson and systematized by Weber. Even capillary effects, contact phenomena par excellence, were subjected by Laplace and by Gauss to the principle of action at a distance.[91]

[91] Cf. Mary B. Hesse, *Forces and Fields: The Concept of Action at a Distance in the History of Physics*, (Thomas Nelson, London, 1961).

CHAPTER 8

THE CONCEPT OF FORCE IN THEOLOGICAL INTERPRETATIONS OF NEWTONIAN MECHANICS

It appears that Newton personally was convinced that gravitational force might eventually be explained, mechanically or nonmechanically, by more profound and more fundamental processes and conceptions. Yet, as indicated in the last chapter, nowhere in his scientific writings does he take a stand on this question. Newtonian physics conceives force and gravitation as an irreducible fact of experience and affords no information on the metaphysical nature of force and gravitation. *Hypotheses non fingit.*

Newton's disciples, followers, and commentators, however, viewed the situation in a different light. Cotes's preface to the second edition of the *Principia* (1713) contributed, though it seems unintentionally, much to the idea that gravitation was to be regarded as an action at a distance and gravitational force as innate in matter. In his enthusiastic effort to acclaim the great explanatory power of Newton's theory of gravitation, which accounts for the motion of the planets as well as of the comets, as compared with the Cartesian vortex theory, which faces insurmountable difficulties in this respect, Cotes describes gravitation loosely as "the nature of gravity in earthly bodies,"[1]

[1] *Sir Isaac Newton's mathematical principles of natural philosophy,* trans. Andrew Motte, rev. Florian Cajori (University of California Press, Berkeley, 1934), p. xxi. (Quoted hereafter as *Principles.*)

as having "a place among the primary qualities of all bodies,"[2] and so on. Although Cotes makes it quite clear that gravity may be regarded as a primitive concept in Newton's conceptual scheme, expressions like these encouraged early metaphysical speculation about the nature of gravitational force.

Still, under the influence of Cartesian physics as the geometry of the extended impenetrable, and because of its aversion to inherent qualities or occult properties, action at a distance was viewed by many theoreticians of the early eighteenth century as a disguise of occult quality and was criticized as incomprehensible. The only way to reconcile this new and immensely successful notion with traditional ideas was to supply it with a metaphysico-theological foundation and to assimilate it into the Neo-Platonic body of doctrines. Force and gravitation were thus conceived as manifestations, *par excellence,* of divine omnipresence and omnipotence. In fact, we shall soon point out in detail how Neo-Platonic thought, still vigorously active in England at the time of Newton, was well prepared for this process of assimilation, so well indeed that Newton himself, an ardent student of theological literature, could almost be classified, as far as his extrascientific conceptions are concerned, as imbued with this spirit. But before discussing this school of thought and its metaphysico-theological interpretation of the concept of force, let us point out that Newtonian physics, together with Leibnizian monadology, formed the basis also of a different school of thought, perhaps best characterized by Rogerius Josephus Boscovich, for whom force was the ultimate element of reality. For the sake of completeness and clarification it should be mentioned in this context that a third system of opinions also claimed Newtonian physics as its point of departure by early in the eighteenth century; according to this system, force is but a relational concept, describing in a convenient manner empirical and measurable relations among

[2] *Principles,* p. xxvi.

perceptible phenomena. That Newton's physics and his theory of gravitation in particular serve as the scientific foundation of fundamentally divergent metaphysical, epistemological, and methodological trends and currents of thought should not surprise us too much if we recall that only a few decades ago, and to some extent still today, Einstein's theory of relativity was in a similar position: it was proclaimed by the Neo-Kantian school as idealistic, by positivistic exponents as operational, and so on.[3] There are not only metaphysical foundations of physics, but also physical foundations of metaphysics. For science and philosophy are, at least genetically, interrelated and they exert mutual influence upon each other.

Let us now discuss the first of these three schools of thought that adopted Newtonian gravitation as the scientific manifesto of their doctrines. It is the Cambridge school of Neo-Platonic thought and its later proponents of natural religion among the commentators of Newton. In order to clarify the position of Newtonian physics, thus interpreted, in relation to Neo-Platonic thought, it would be advisable to review briefly the history of English Neo-Platonism up to the time of Newton.

We have shown in a previous chapter how the Italian Platonic Renaissance influenced Cusanus' and Copernicus' conceptions of force. Italian humanism, originally in harmonious accord with the church and allied to the forces of religion, slowly changed its position; first adopting a deliberate skepticism toward the traditional values of religious faith, it soon challenged the dogma and came out into open opposition against the Christian way of life and Christian ethics.

This intellectual revolution, culminating perhaps in Giordano Bruno, had its beginning as early as the early fifteenth century, when Laurentius Valla, a papal secretary, declared himself an

[3] See, for instance, the discussion by H. W. Carr, T. P. Nunn, A. N. Whitehead, and Dorothy Wrinch in *Proceedings of the Aristotelian Society* [n.s.] *22*, 123 (1922).

opponent of tradition by his criticism of scholastic method and aims.[4] Already his first philosophical publication, *De voluptate*, printed in 1431, favoring ancient Epicurean ideals of life, placed him in antagonism to the church. English humanism had a different course. Its criticism was also directed against scholastic insipidity and traditional forms of theological learning, yet it never opposed religion. On the contrary, differentiating between religion and scholasticism, it fought for the former in its battle against the latter. William Grocyn and Thomas Linacre, inspired by Italian Neo-Platonic thought while studying in Padua, transmitted these ideas to Oxford and London at the end of the fifteenth century without stirring any controversy with the church. Moreover, the ground was thus prepared for the reaction against the materialism and mechanism of philosophical teachings such as Hobbes's philosophy of nature, a process that had its climax in Ralph Cudworth's and Henry More's writings. Cudworth's conception of "Plastick Nature" is a variation of the Neo-Platonic concept of a world soul, but with an important reservation: the universe is not conceived as activated by forces working from without, but by a forming principle from within; but this principle, this "Plastick Nature," does not operate consciously, as the world soul was supposed to do, and consequently cannot be identified with God's being; it serves rather as an unconscious intermediary, a medium, under the control of the highest intelligence which guides the cosmos. Ralph Cudworth, in his *Intellectual system of the universe*,[5] spoke of matter as being passive, since its impenetrability and extension do not yet imply any form of activity, and as endowed from without with an incorporeal essence of self-activity, forces that pursue ends without deliberation. In opposition to Descartes's dualism of mind and matter, Cudworth's construction of this

[4] Laurentius Valla, *Dialecticae disputationes contra Aristotelicos* (Venice, 1499).

[5] Ralph Cudworth, *The true intellectual system of the universe* (London, 1687).

modified dualism was to escape mechanism without falling into occasionalism.

> Wherefore, since neither all things are produced fortuitously, or by the unguided mechanism of matter, nor God himself may reasonably be thought to do all things immediately and miraculously; it may well be concluded, that there is a plastic nature under him, which, as an inferior and subordinate instrument, doth drudgingly execute that part of his Providence, which consists in the regular and orderly motion of matter.[6]

For Cudworth, force is thus ultimately an obscure impulsion, a blind activity of nature, and yet a divine institution in the service of God. In fact, it is not surprising that Cudworth's conception of force shows some resemblance to that of Philo Judaeus; "his plastic natures are a sort of 'third man' designed to bridge the gap between God and matter, mind and body."[7]

In his opposition to Descartes's mechanism and Hobbes's materialism Cudworth was joined by Henry More, one of the most famous Cambridge Platonists. More, who died in the year in which the *Principia* was published, had none the less participated in the discussion on gravitation, then in vogue. In his effort to prove the existence of immaterial spirituality, More enlists not only space, as we have shown elsewhere,[8] but also gravity for this purpose. In his *Antidote against atheism* an important chapter carries the title: "A demonstration from the phenomenon of gravity, that there is a principle distinct from matter."[9] Similar to Cudworth's modified notion of the Neo-Platonic world soul, the spirit of nature is

> a substance incorporeal, but without sense and animadversion, pervading the whole matter of the universe, and exercising a plastic power therein, according to the sundry predispositions and occasions in the parts it

[6] *Ibid.* (first American ed., New York, 1837), vol. 1, p. 213.

[7] J. A. Passmore, *Ralph Cudworth, an interpretation* (Cambridge University Press, Cambridge, 1951), p. 28.

[8] Max Jammer, *Concepts of space, the history of theories of space in physics* (Harvard University Press, Cambridge, 1954), pp. 38–46.

[9] Henry More, *An antidote against atheism* (London, 1653), book II, chap. II, art. 7.

works upon, raising such phenomena in the world, by directing the parts of the matter, and their motion, as cannot be resolved into mere mechanical powers.[10]

More was unhesitatingly convinced — and this is the irrevocable foundation of his whole intellectual outlook — that any mechanical explanation of natural phenomena is insufficient, or, as he wrote in a letter to Boyle, "the phenomena of the world cannot be solved mechanically, but there is the necessity of the assistance of a substance distinct from matter, that is, of a spirit, or being incorporeal."[11] This immaterial substance, "penetrable and indiscerptible," as he calls it in *The immortality of the soul,* fulfills God's will in the material world, like Cudworth's plastic nature, by being endowed by the highest intelligence with "self-motion, self-contraction, and dilatation."[12] It is this intrinsic elastic capability of contraction and dilatation that accounts for the motion of material substances. Were it not for this spiritual nature, a stone let loose above the surface of the earth would never fall in a straight line in the direction toward the earth's center, but would be carried along by Cartesian vortices or fly off on the tangent to the surface of the earth, as More explained in his *Enchiridion metaphysicum, sive de rebus Incorporeis."*[13] Since, however, this spirit of nature is but "the vicarious power of God upon matter," as expounded in the preface to *The immortality of the soul,* God is the ultimate originator of motion and of gravity. Thus Henry More comes to his final conclusion: "Nam sic mobilia omnia moventur a Deo."[14]

Force as well as space have their origin in God. Extension

[10] Henry More, *The immortality of the soul* (London, 1659; ed. of 1718), book III, chap. 12, part I.

[11] *The works of the Hon. Robert Boyle,* ed. T. Birch (London, new ed., 1772), vol. 6, p. 513.

[12] More, *The immortality of the soul,* book I, chap. V; also book I, chap. II, art. 11, 12.

[13] Henry More, *Enchiridion metaphysicum, sive de rebus incorporeis* (London, 1671), chap. 2, art. 14.

[14] *Ibid.* chap. 28, part I, art. 3.

and motion are but manifestations of God's omnipresence and omnipotence.

> Onely that vitality,
> That doth extend this great Universall,
> And move th'inert Materiality
> Of great and little worlds, that keep in memory.[15]

More exerted a strong influence on Locke, Newton, and Clarke and through them on the eighteenth century in general. It is, however, interesting to note that, contrary to his concept of space, Newton's concept of force and gravitation shows only a faint reverberation of More's ideas. While More's notion of force is an essentially theistic conception, Newton's extrascientific view accords more with the deistic conception which recognizes a divine intervention in the past as the first cause that started the motion of the planets and of the comets, and, apart from possibly regulative interferences at rare times, determined only their course to their final ends. It is perhaps again in the *Opticks* that Newton alludes to More's conceptions, when he says:

> The main business of natural philosophy is to argue from phenomena without feigning hypotheses, and to deduce causes from effects, till we come to the very first cause, which certainly is not mechanical; and not only to unfold the mechanism of the world, but chiefly to resolve these and such like questions. What is there in places almost empty of matter, and whence is it that the sun and planets gravitate towards one another, without dense matter between them? Whence is it that nature doth nothing in vain; and whence arises all that order and beauty which we see in the world? To what end are comets, and whence is it that planets move all one and the same way in orbs concentric, while comets move all manner of ways in orbs very eccentric, and what hinders the fixed stars from falling upon one another? . . . How do the motions of the body follow from the will, and whence is the instinct in animals? . . . And these things being rightly dispatched, does it not appear from phenomena that there is a being incorporeal, living, intelligent, omnipresent, who, in infinite space, as it were in his sensory, sees the things themselves intimately, and thoroughly perceives them; and comprehends them wholly by their immediate presence to himself? [16]

[15] Henry More, "Psychathanasia," book II, canto I, p. 108, in *Psychodia platonica* (London, 1642).

[16] Isaac Newton, *Opticks* (Dover, New York, 1952), query 28, p. 369.

In these words a strong emphasis on space as God's omnipresence is recognizable, but very little is said about force as an immediate divine operation. More's philosophy, in short, may have had, even as far as the concept of force is concerned, a more or less emotional effect upon Newton, but certainly had no considerable influence on the conceptual formulation of this notion.

We are now in a position to assess Böhme's indirect influence upon Newton, a question that was left for this occasion after we rejected a direct influence in Chapter 7. There is no doubt that Henry More studied Böhme extensively. In 1670 More published a detailed critique of Böhme's mystical writings in his *Philosophiae teutonicae censura*.[17] But already earlier, in 1656, in his *Enthusiasmus triumphatus*,[18] More refers to Böhme's conceptions in natural philosophy. Popp[19] even suggests that More's estrangement from Descartes originated from his reading Böhme's works at some time after his correspondence with Descartes and before the composition of his *Enchiridion metaphysicum*. This statement perhaps underestimates the decisive influence of English Neo-Platonic thought upon More at that time and his cabalistic studies. But whatever the case, it is abundantly clear that even this indirect influence of Böhme upon Newton via More had little impact as far as Newton's conceptions of force and gravitation are concerned.

More's spiritual exposition of force and gravity does not yet form the final stage in the development, started by the early English humanists, of interpreting natural phenomena in a theological manner for the sake of promoting religion. In fact, it was Newton's conception of force and gravitation, or rather ideas which could easily be associated therewith, that gave a new

[17] Henry More, *Philosophiae teutonicae censura* (1670), in *Opera omnia* (London, 1679).

[18] Henry More, *Enthusiasmus triumphatus: or a brief discourse of the nature, causes, kinds and cure of enthusiasm* (1656), in *A collection of several philosophical writings of Dr. Henry More* (London, Cambridge, 1662).

[19] K. R. Popp, *Jakob Böhme und Isaac Newton* (inaugural dissertation; Gerhardt, Leipzig, 1935), p. 42.

impetus to theological speculation. Indeed, it did not take long for Newtonian principles to be thus interpreted, assimilated and made use of for apologetic purposes. Richard Bentley, who, we have seen, was one of Newton's closest friends, interpreted Newton's physics, and his conception of force and gravitation in particular, for his stand against atheism and deism. William Whiston[20] advocated a doctrine that unites Newtonian attraction with the principles of natural religion and the Bible. Typical of these attempts is the title of William Derham's Boyle lecture, *Physico-theology or a demonstration of the being and attributes of God from his works of creation.*[21] In another work, called *Astrotheology,* a book that received great publicity and was soon translated into several foreign languages, Derham speaks of gravitation as "an active quality, impressed on matter by the great Creator,"[22] and he dedicates long chapters to a detailed exposition of the great advantages of gravitation and its beneficial influence for the conservation and integrity of matter and life. Samuel Horsley, the editor of Newton's collected works, emphasizes the divine origin of gravity:

> Appetentiam illam et fugam non corporibus ipsis naturalem statuit; non cum Epicuro aeternam, eandem tamen fortuitam; non a Veterum quorundam Amore et Odio, Juniorum Sympathia et Antipathia profectam; sed a causa quadam incorporea, quae, ut sensus nostras maxime lateat, intellectum tamen et mentis acum haud fugit. A Deo utique haec omnia proficisci palam vociferatus est.[23]

Attraction and repulsion are not to be conceived as natural properties of bodies, or eternally intrinsic in matter, as claimed by Epicurus, nor as love and strife of the ancients, nor sympathy and antipathy of the moderns.

[20] William Whiston, *A new theory of the earth* (London, 1696).

[21] William Derham, *Physico-theology or a demonstration of the being and attributes of God from his works of creation* (London, 1716).

[22] William Derham, *Astrotheology or a demonstration of the being and attributes of God from the survey of the heavens* (Fifth ed., London, 1726), book V, p. 135.

[23] Samuel Horsley, *Newtoni opera* (London, 1779–1785), vol. I, "Dedicatio ad Regem," p. iii.

Similarly, Andrew Baxter, the great Scottish metaphysician, who became very influential upon Joseph Priestley and Priestley's friend John Michell, contends that matter itself is absolutely inactive and passive; force and gravitation are impressed upon matter by God. In his *Inquiry into the human soul* Baxter contrasts inertia, which he regards as the essential character of all matter, to every change of its present state, with active power of force, showing that these two are incompatible, unless an immaterial mover or a universal providence is operative for their reconciliation.[24]

This list of theological interpretations of Newton's conceptions of force and gravitation could be extended considerably, yet without adding much originality of ideas. For further information the interested reader is referred to Hélène Metzger's *Attraction universelle et religion naturelle chez quelques commentateurs anglais de Newton.*[25]

This movement, however, was not confined to English commentators alone. An interesting example of a theistic interpretation of Newton's conception of attraction is P. L. Moreau de Maupertuis's article "Sur les loix de l'attraction," [26] in which this famous proponent of the principle of the least action considers the possibility of laws of attraction other than the inverse-square law. By a conclusive mathematical demonstration Maupertuis tries to show that God's choice of the inverse-square law for the transmission of gravitational forces endows nature with a harmonious unity that it would not exhibit had God selected a different law of propagation.

Si donc le Créateur et l'Ordonnateur des choses avoit voulu établir quelque loi d'attraction dans la matière, on vit et l'on va voir encore mieux par la théorie suivante, que toutes les loix n'auroient pas dû lui

[24] Andrew Baxter, *An inquiry into the human soul wherein the immateriality of the soul is evinced from the principles of reason and philosophy* (London, 1737), sec. 1.

[25] Hélène Metzger, *Attraction universelle et religion naturelle chez quelques commentateurs anglais de Newton* (Hermann, Paris, 1938).

[26] Pierre Louis Moreau de Maupertuis, "Sur les loix de l'attraction," *Mémoires de l'Académie Royale* (Paris, 1732), pp. 343–362.

paroître égales. En effet, s'il avoit fait un choix il y aurait eu sans doute des raisons pour ce choix.[27]

In his demonstration Maupertuis refers to a theorem from Newtonian potential theory that states that only spherical bodies exert equal forces in all directions. The only definite point of reference for the determination of distance in this case is the center of the sphere. If, now, the Creator aimed at a uniformity of action in matter, the law should apply equally for a mass point as well as for an extended sphere. Under this restriction only two alternatives are possible: either attraction increases linearly with distance or it decreases inversely with the square of the distance. Since an augmentation of the effect with distance is inconceivable, Maupertuis argues, Newton's inverse-square law remains as the only logical possibility.

[27] *Ibid.*, p. 348

CHAPTER 9

DYNAMISM: LEIBNIZ, BOSCOVICH, KANT, SPENCER

Cudworth's, More's, and Newton's concepts of force, which formed the foundation of the theological interpretations discussed in the last chapter, originated from a reaction to the Cartesian view of geometrized physics which recognized only extension and impenetrability. Leibniz's view had a similar origin; in fact, Leibniz himself recognized that there were many points of contact between his own concepts and those of the Cambridge Platonists. Yet, as we shall see, Leibniz rejected the doctrine of "plastic nature" not because of its contents but because of its method. Although it is almost axiomatic for Leibniz's philosophical and scientific reasoning that the physico-mechanical and the theologico-teleological conceptions of the world should not exclude each other, he nevertheless rejects any spiritual principles for the explanation of physical phenomena, which, he contends, should be accounted for only and exclusively by the principles of corpuscular philosophy, that is, by the physical sciences and their mathematical method of reasoning.

With Leibniz the concept of force undergoes a radical change of meaning: from a mechanical mode of operation it becomes a principle of almost vitalistic activity. Strictly speaking, Leibniz's concept of force is what we call today kinetic energy, but conceived as inherent in matter and representing the innermost nature of matter. Because of the great importance attached to this concept in Leibniz's metaphysical and scientific outlook, he may rightly be considered as the first proponent of modern dynamism in natural science.

In 1669, in a letter to Jacob Thomasius, Leibniz still adheres to the Cartesian conception of corporeal objects as consisting in mere extension. Yet very soon afterward he revised his view on the constitution of matter. In his autobiographical letter to Rémond de Montmort, in 1714, Leibniz describes his intellectual development in the following words:

> After I had left the elementary school, I fell in with the modern philosophers, and I recall that I was walking alone in a little grove called the Rosenthal, not far from Leipzig, at the age of fifteen years, in order to make up my mind whether I should adopt the doctrine of substantial forms. The theory of mechanism, however, won finally the upper hand with me and led me to mathematics. Yet, in my search for the ultimate grounds of mechanism and of the laws of motion, I came back to metaphysics and the doctrine of entelechies.[1]

Recognizing that a consequent elaboration of Descartes's conception must inevitably lead to some kind of Spinozism, Leibniz emphasized in his *Hypothesis physicae nova,* in March 1671, that the purely kinetic point of view, advanced by Descartes, is too one-sided for a comprehensive understanding of natural phenomena and must be supplemented by a dynamic principle; the essence of matter cannot consist exclusively of extension and motion.

Meanwhile Francis Glisson, an English physician and philosopher intellectually related to the Cambridge Platonists, published his *Treatise on the energetic nature of matter,*[2] in which Leibniz found, to say the least, support and confirmation for his new view on the nature of matter. During his service as librarian and archivist at the ducal library of the House of Brunswick in Hanover (and perhaps even earlier at his father's

[1] G. W. Leibniz, *Opera philosophica quae extant, latina, gallica, germanica omnia,* ed. J. E. Erdman (Berlin, 1840), p. 701.

[2] Francis Glisson, *Tractatus de natura substantiae energetica seu de vita naturae, ejusque tribus primis facultatibus* (London, 1672).

library) he hit upon Cudworth's and More's works, to which he often refers in his later writings. Encouraged by these works, Leibniz formulated his new conception of matter, first tentatively in a series of articles in the *Acta Eruditorum,* and later more systematically, as far as one can speak of a systematic exposition in Leibniz's writings at all, in his *Monadology.* One of his first clear statements in this respect is found in his *De primae philosophiae emendatione et de notione substantiae,* in which he declares:

> To give a foretaste of my conceptions, it is sufficient for me to explain that the notion of force or virtue which the Germans call "Krafft," the French "la force," and for the exposition of which I have designed a special science of dynamics, adds much to the clear understanding of the concept of substance. For active force differs from the concept of bare power familiar to the Scholastics, in that this potentiality or faculty of the schools is nothing but a possibility ready to act, which nevertheless needs an external excitation or stimulus, as it were, in order to pass into action. But active force contains a certain activity or entelechy and is midway between the faculty of acting and the action itself; it includes effort and thus passes into operation by itself, without any auxiliary, but with only the removal of impediments.[3]

These remarks are of great interest for the history of physics, not only because they declare an independent ontological status of the Leibnizian concept of force, but also because they proclaim a new special science — called, perhaps for the first time, dynamics — as the study and investigation of its manifestations in nature.

Leibniz was not only a theologian and metaphysician, but also

[3] "Cujus rei ut aliquem gustum dem, dicam interim, notionem virium seu virtutis (quam Germani vocant 'Krafft,' Galli 'la force') cui ego explicandae peculiarem Dynamices scientiam destinavi, plurimum lucis afferre ad veram notionem substantiae intelligendam. Differt enim vis activa a potentia nuda vulgo scholis cognita, quod potentia activa Scholasticorum, seu facultas, nihil aliud est quam propinqua agendi possibilitas, quae tamen aliena excitatione, et velut stimulo indiget, ut in actum transferatur. Sed vis activa actum quendam sive entelecheian continet, atque inter facultatem agendi actionemque ipsam media est, et conatum involvit; atque ita per se ipsam in operationem fertur; nec auxiliis indiget, sed sola sublatione impedimenti." *Gothofredi Guillelmi Leibnitii opera omnia,* ed Ludovicus Dutens (Geneva, 1768), vol. 3, pp. 19–20.

a physicist and mathematician. No wonder, therefore, that he was searching for qualitative and quantitative determinations of his concept of force within the conceptual scheme of physics proper. For qualitative demonstrations he refers to the principle of inertia. Natural inertia, he contends, cannot be accounted for by the mere conception of extension. With him, inertia becomes a real *vis insita* in a dynamical sense. "Pour prouver que la nature du corps ne consiste pas dans l'étendue," he declares, "je m'étois servi d'un argument expliqué dans le Journal des Sçavans du 18. Juin 1691, dont le fondement est, qu'on ne sçauroit rendre raison par la seule étendue de l'inertie naturelle des corps, c'est-à-dire, de ce qui fait que la matière résiste au mouvement." [4] For Leibniz, a moving body is different from a body at rest. Its motion is not merely a successive occupancy of different places in space, it is a state of motion at each separate moment. This state of continual change of place involves some effort. As, however, the very principle of inertia excludes an external influence for the continuation of this motion with constant speed, this effort must be the outcome of an inherent force or activity. Inertia as the principle of the continuation of motion is thus a proof of the existence of an activity inherent in the moving body. Moreover, inertia as the principle of resistance, to be overcome by the moving forces, must be of the same category as these, that is, it must be a force.

One point only in these conclusions needs further elucidation. It may be objected that Leibniz's conception of motion as explained above is merely a variation of the old impetus theory and thus implies no original ideas. That this interpretation is wrong can be easily understood if we remember that Leibniz's notion of motion is the result of his view of space as a relational concept. Since space is relational, he contends, that which is real and absolute in motion does not consist in what is purely

[4] "Extrait d'une lettre de Mr. Leibniz pour soutenir ce qu'il y a de lui dans le Journal des Sçavans du 18. Juin 1691," *ibid.*, vol. 2, p. 236.

mathematical, such as change of neighborhood or localization: it is force itself.

If motion is nothing but change of contact or immediate vicinity, it follows that it can never be determined which thing is moved. For as in astronomy the same phenomena are represented in different hypotheses, so it is always permissible to ascribe real motion to either one or the other of those bodies which change among themselves vicinity or situation . . . Therefore in order that a thing can be said to be moved, we require not only that it change its situation in respect to others, but also that the cause of change, the force or action, be in itself.[5]

Thus, relativity of space and absoluteness of motion led Leibniz to the existence of force, in contrast to Newton, for whom absoluteness of force proved the absoluteness of motion and, consequently, the absoluteness of space.[6] In his monadology Leibniz resumes his former conclusions and relegates resistance or the force of inertia to the so-called *prima materia,* the final substratum of infinitely divisible matter. Furthermore, he even accounts for the extension of matter by this principle of resistance, an idea which Kant later discusses in much detail. Finally, since space is extended but obviously not impenetrable, impenetrability cannot be reduced to extension, and consequently also not to the inherent force of inertia. Primary matter, therefore, in Leibniz's view, has two ultimate qualities: dynamical inertia (a paradoxical combination of terms!) and impenetrability (or, as Leibniz prefers to call it, *antitypia*).

Let us now turn to Leibniz's quantitative determination of his

[5] G. W. Leibniz, "Animadversions on Descartes' principles of philosophy" (1692), in *The philosophical works of Leibnitz,* ed. G. M. Duncan (second ed., New Haven, 1908), p. 62. Cf. also Leibniz's "On nature in itself; or on the force residing in created things, and their actions" (1698): "For a body is not only at the actual moment of its motion in a place commensurate to it, but it has also a tendency or effort to change its place so that the succeeding state follows of itself from the present by the force of nature; otherwise at the actual moment, and hence at any moment, a body *A*, which is in motion, could in no wise differ from a body *B*, which is at rest." *Ibid.,* p. 129.

[6] See Max Jammer, *Concepts of Space* (Harvard University Press, Cambridge, 1954), pp. 103–104.

concept of force. Here, again, it should be remarked that Leibniz does not yet have a clear quantitative conception of mass, which he calls sometimes *moles*, sometimes *corpus*, sometimes by other names. Only in his article of 1795, to be mentioned later, does he use the term *massa*, which he most probably borrowed from Newton. However, for the sake of a clear and concise rendition of the mathematical implications in Leibniz's writings we shall interpret consistently his *moles*, etc., as we did in the case of Descartes, Huygens, and Galileo, as mass.

In an interesting article, published in 1686 in the *Acta Eruditorum* under the title "A short demonstration of a remarkable error of Descartes and others, concerning the natural law by which they think that the Creator always preserves the same quantity of motion; by which, however, the science of mechanics is totally perverted,"[7] Leibniz discusses the quantitative aspects of force. He contends that the Cartesian quantity of motion, that is, the product of mass and velocity, is not the general measure of force. He explains this misconception as arising from the fact that in the five ordinary machines speed and mass "compensate each other."[8] Hence, Leibniz continues, the Cartesians jumped to the conclusion that it is this quantity which is conserved in the universe (and constitutes therefore the mathematical measure of force). To expound the correct measure of force, Leibniz starts with the assertion that the same force is required to lift a mass m_a of 1 lb to a height h_a of 4 ft as a mass m_b of 4 lb to a height h_b of 1 ft. The same force will be present if these bodies fall through these distances respectively. Now, as shown by Galileo,

[7] "Brevis demonstratio Erroris memorabilis Cartesii, et aliorum circa legem naturalem, secundum quam volunt a Deo eandem semper quantitatem motus conservari; qua et in re mechanica abutuntur," *ex Actis Erudit.* (Lips. ann. 1686), in *Opera omnia,* vol. 3, p. 180.

[8] "Complures Mathematici cum videant in quinque machinis vulgaribus celeritatem, et moles inter se compensari, generaliter vim motricem aestimant a quantitate motus, sive producto ex multiplicatione corporis in celeritatem suam." *Ibid.* The principle referred to is essentially the medieval principle of virtual work: the loads are inversely proportional to the velocities of displacement.

the speed v_a acquired by the first body falling 4 ft is twice the speed v_b of the second body which descended only 1 ft:

$$v_a = 2v_b.$$

Under the natural assumption that the force is proportional to the mass m, the following equation holds:

$$\text{force} = mf(v),$$

in which $f(v)$ denotes a function of the speed yet to be determined. As the two forces are equal, we have

$$m_a f(v_a) = m_b f(v_b).$$

But $m_b = 4m_a$ and $v_a = 2v_b$. Therefore

$$m_a f(2v_b) = 4m_a f(v_b) \quad \text{or} \quad f(2v_b) = 4 f(v_b).$$

The last equation, Leibniz concludes, shows that the function $f(v)$ must be a quadratic function of its argument, namely, v^2. What is conserved, and what is the measure of force, is therefore

$$mv^2.$$

"Eodem modo generaliter colligitur, vires aequalium corporum esse ut quadrata celeritatum, et proinde vires corporum in universum in ratione composita ex corporum simplice, et celeritatum duplicata." [9] This is Leibniz's mathematical definition of force in its state of action, quoted from his "Specimen dynamicum" (1695), to which reference has been made above. In this article Leibniz also introduces a new term for this expression, calling it *vis viva*[10] in contrast to *vis mortua*, an expression coined by Galileo to designate what we call today "pressure" or "tension" in the unscientific meaning of the word, that is, force not associated with motion.

[9] "Specimen dynamicum pro admirandis naturae legibus circa corporum vires, et mutuas actiones detegendis, et ad suas causas revocandis," ex *Actis Erudit.* (Lips, ann. 1695), in *Opera omnia,* vol. 3, p. 323.

[10] "Hinc 'vis' duplex: alia elementaris, quam et mortuam appello, quia in ea nondum existit motus, sed tantum sollicitatio ad motum, qualis est globi in tubo, aut lapidis in funda, etiam dum adhuc vinculo tenatur; alia vero vis ordinaria est, cum motu actuali conjuncta, quam voco vivam." *Ibid.,* p. 318.

Leibniz's article of 1686 started the famous controversy between the Leibnizians and the Cartesians as to whether the appropriate measure of force is the "quantity of motion" mv or the "*vis viva*" mv^2, a controversy that engaged almost all leading mathematicians and physicists for nearly half a century. The Abbé de Catelan, P. Jean Simon Mazière, Colin Maclaurin, James Stirling, Samuel Clarke, and Jean Jacques d'Ortons de Mairan associated themselves with the Cartesian conception, while John Bernoulli, William Jacob s'Gravesande, Christian Wolf, Georg Bernhard Bulfinger, and Samuel König defended the Leibnizian point of view in this dispute. Without going into details and discussing the various arguments of the participants in this discussion,[11] it may be stated that it was essentially a mere battle of words, since the disputants discussed different concepts under the same name. In modern terminology the issue may be stated as follows. When comparing two constant forces f and F, claimed the Cartesians, let them act for a given time interval t; the ratio of these two forces is then given by

$$\frac{f}{F} = \frac{ma}{MA} = \frac{mat}{MAt} = \frac{mv}{MV},$$

that is, the forces are proportional to their corresponding quantities of motion. The Leibnizians, on the other hand, said, let the forces act through a given distance s. Since, according to Galileo

$$v = \sqrt{(2as)} \quad \text{and} \quad V = \sqrt{(2As)},$$

their ratio turns out to be

$$\frac{f}{F} = \frac{ma}{MA} = \frac{mas}{MAs} = \frac{\frac{1}{2}mv^2}{\frac{1}{2}MV^2} = \frac{mv^2}{MV^2},$$

that is, the forces are as their corresponding *vires vivae* or "living forces." It is clear that Descartes's approach gives a correct measure for forces acting for equal times, whereas Leibniz's *vis viva*

[11] For further information see, for example, René Dugas, *Histoire de la mécanique* (Éditions Dunod, Paris, and Éditions du Griffon, Neuchatel, 1950), pp. 225ff. Cf. also the same author's *La mécanique au XVIIe siècle* (Éditions Dunod, Paris, and Éditions du Griffon, Neuchatel, 1954), pp. 477ff.

is a correct measure for forces acting through the same distance.

What Leibniz really determined was, of course, the efficacy of the force as manifested in the moving body. One half of the *vis viva,* that is, $\frac{1}{2}mv^2$, was later called by Jean Baptista Charles Joseph Bélanger "the living power"[12] and is known today as "kinetic energy." Consequently, it is our concept of energy that was referred to by Leibniz as "force."[13] We have seen already in Chapter 1 how unfortunate this terminology was. Its detrimental effect on the subsequent development of the concepts of force and energy in mechanics was increased by the fact that Newton's *Principia,* the bible of classical mechanics, did not discuss any energetic concepts or theorems. It is, however, not to be doubted that Newton and Leibniz, as well as Huygens, were groping with an intuitive conception of an energetic principle which they believed lay at the foundation of all physical phenomena in nature.

It is always dangerous to interpret intuitive notions in modern terminology; it is too easy to read into them more than they contain. Yet, in the case mentioned, documentary evidence makes it abundantly clear that an anticipation of the principle of conservation of energy was at the bottom of the issue. Of course, the term "energy" was not mentioned, nor was the term "work"[14]

[12] Gustave Gaspard Coriolis introduced the factor $\frac{1}{2}$ in Leibniz's *vis viva* for the sake of mathematical convenience, in his *Calcul de l'effet de machines, ou considérations sur l'emploi des moteurs et sur leur évaluation* (Paris, 1829).

[13] That Leibniz was aware of this is shown in his letters to Arnauld (1690): "As regards physics, it is necessary to understand the nature of force, a thing entirely different from motion, which is something more relative. That this force is to be measured by the quantity of effect." *The philosophical works of Leibnitz,* p. 40.

[14] With reference to our remark about the danger of a too liberal interpretation of vague expressions, let us quote two passages bearing upon the history of the concept of work. Philipp Lenard in his book *Great men of science* (Macmillan, New York, 1933), p. 10, says about Leonardo da Vinci: "He was probably the first to clearly distinguish the concept of work from that of force, inasmuch as he remarks that work implies a distance moved in the direction of the force." I. B. Hart, on the other hand, in his book *The mechanical investigations of Leonardo da Vinci* (London, 1925), p. 93, states: "It need scarcely be said at the outset that da Vinci never made use of the term 'work' in the modern technical sense. Nor did he have any clear ideas as to the importance of the product of force and distance." Upon a perusal of

or its equivalent in Latin or French. But when Huygens wrote to Leibniz in 1692 about certain papers submitted by him to the Royal Society "dans lesquelles j'emploiay avec autre chose cette conservatio virium aequalium et la deduction au mouvement perpetuel, c'est à dire à l'impossible,"[15] his quest for a quantity invariant in impact phenomena together with his statement about the impossibility of a *perpetuum mobile* leaves no doubt that what he had in mind was a principle of conservation of energy. His search for such a principle was complicated by the fact that he believed atoms to be inelastic; on the other hand, he was unable to account for the loss of motion (or energy) in inelastic impacts.

Leibniz, on the contrary, conceived atoms as elastic, which seemed to him more in accordance with his monadology (a point of great importance for the further development of Leibnizian dynamism into theories of force centers). This assumption also made it possible for Leibniz to maintain consistently that "force" is conserved in the universe. For in the case of elastic collisions, Huygens, in an article in the *Journal des Sçavants*[16] of 1699 and in another paper,[17] published in the *Philosophical Transactions of the Royal Society*, of the same year, had shown that the sum of the products of the masses and the squares of their respective velocities, before the collision, is equal to the corresponding expression after the collision. In other words, the *vis viva* is not diminished in an elastic collision. But how does Leibniz account for his assertion of the constancy of motion in the case of inelastic collisions? In reality, he says, inelastic collisions are only macroscopic phenomena; the loss of "force" is only apparent. He

---

the published manuscripts of Leonardo we were unable to substantiate Lenard's statement. It may likewise be contended that even Leibniz had no clear conception of the notion of work, although in his derivation of *vis viva* the concept is made use of implicitly. It was Coriolis who finally coined the term "work" for the product of force and distance.

[15] Huygens's letter of July 11, 1692, in *Oeuvres complètes*, vol. 10 (The Hague, 1905), p. 303.

[16] *Journal des Sçavants 2*, 534 (1699).

[17] *Philosophical Transactions of the Royal Society of London 4*, 928 (1699).

objects, that two soft or unelastick bodies meeting together, lose some of their force . . . 'tis true, their wholes lose it with respect to their total motion; but their parts receive it, being shaken by the force of the concourse. And therefore that loss of force is only apparent. The forces are not destroyed, but scattered among the small parts. The bodies do not lose their forces; but the case here is the same, as when men change great money into small.[18]

It is obvious that Leibniz's explanation is a precursor of the modern principle of transformation of energy which contends that in an inelastic collision the decrease of kinetic energy is accounted for by the quantity of heat caused by the impact, that is, ultimately by the increase of molecular energy. Leibniz advanced his theory as an argument against Newton's followers who "have also a very odd opinion concerning the work of God. According to their doctrine, God Almighty wants to wind up his watch from time to time: otherwise it would cease to move." [19] Since the Newtonians were of the opinion that "force" decreases constantly in inelastic collisions, new "forces" have to be supplied from time to time by God to the universe, which otherwise would come to a complete standstill. Clarke, in defense of Newton's position, discusses the interesting question whether such injection of "new forces" constitutes a natural or a supernatural, miraculous process. Leibniz opposes such intervention as incompatible with God's power and foresight. He says:

He [God] had not, it seems, sufficient foresight to make it a perpetual motion. Nay, the machine of God's making is so imperfect, according to these gentlemen, that he is obliged to clean it now and then by an extraordinary concourse, and even to mend it as a clockmaker mends his work; who must consequently be so much the more unskillful a workman, as he is oftener obliged to mend his work and to set it right. According to my opinion, the same force and vigour remains always in the world, and only passes from one part of matter to another, agreeably to the laws of nature.[20]

[18] Leibniz's fifth paper in *A collection of papers which passed between the late learned Mr. Leibniz and Dr. Clarke* (London, 1717), p. 253; *The Leibniz-Clarke Correspondence*, ed. H. G. Alexander (Philosophical Library, New York, 1956), p. 87.

[19] *Ibid.*, pp. 3–5; 1956 ed., p. 11.

[20] *Ibid.*, pp. 5–7; 1956 ed., pp. 11–12.

Leibniz's objection to this theistic principle of divine intervention is, consequently, the expression of a deep religious conviction. His physics accords and harmonizes with his theology, yet without mutual interpenetration. As we have mentioned before, spiritual principles cannot be employed for physical explanations nor the converse. Theological convictions can affect physical considerations only so far as they may indicate a certain direction of thought or eliminate assumptions that may ultimately lead to contradictions with the religious credo.[21] We have seen that such was the case with his conception of force as a fundamental constituent of matter, as Descartes's reduction of matter to mere extension would inevitably lead to Spinozism or atheism. In this context it may be mentioned, further, that Leibniz saw a confirmation of his concept of force in his views on the Eucharist, a theologico-metaphysical problem that engaged him and his contemporary Bossuet to a not inconsiderable extent, as the Cartesian concept was for him inconsistent with the principles of transubstantiation or consubstantiation.[22] Leibniz's insistence on the conservation of "force" and its consequent independence of extraneous agencies was of great importance for the further development of dynamism into a theory according to which force was the ultimate element of reality. "Agere est character substantiarum!" These words, which Leibniz wrote in his *Specimen dynamicum,*[23] may serve as the motto of dynamic mechanism.

[21] Compare Leibniz's remarks in *On nature in itself; or on the force residing in created things, and their actions* (1698): ". . . final cause is not only useful to virtue and to piety in ethics and in natural theology, but even in physics it serves to find and to discover hidden truths." *The philosophical works of Leibnitz,* p. 121. For the original text see Leibniz, *De ipsa natura, sive de vi insita actionibusque creaturarum,* in *Opera omnia,* vol. 2, pars II, p. 51.

[22] Jesus Christ's true presence in the "blessed sacraments of the altar" under the appearance of bread and wine would be incompatible with the identification of substance with extension and impenetrability. For Leibniz's views on these questions and their importance for his conception of force see G. E. Guhrauer, *Leibnitz: eine Biographie* (Breslau, 1846), vol. 1, p. 77.

[23] "Agere est character substantiarum, extensioque nil aliud, quam iam praesuppositae nitentis, renitentisque, id est resistentis substantiae continuationem, sive diffusionem, dicit; tantum abest, ut ipsammet substantiam facere possit." (Activity is the characteristic of substances; extension, on the other

"Quod non agit, non existit" ("Whatever does not act, does not exist"). These words characterize Leibniz's dynamism in contrast to the traditional "Operari sequitur esse" ("Action follows existence"). Yet, as Bertrand Russell has pointed out,[24] Leibniz's conceptual scheme of physics only partly conforms to his metaphysical doctrine. Methodological, and probably also personal, reasons led Leibniz to reject Newton's theory of gravitation and in particular its alleged interpretation as action at a distance, an interpretation which by the early eighteenth century was increasingly gaining ground in certain quarters. Leibniz's rejection of action at a distance left him with the alternative of assuming that motion results only from contact, or, as the traditional Latin phrase put it: "Corpus a corpore non moveri, nisi contiguo et modo." Impact, or impulse, as it was often called at that time, was thus for Leibniz the only possible form of mechanistic interaction. Strictly speaking, in accordance with his monadological point of view, no transmission of force was implied, but rather a mutual release of inherent activity. But in spite of this rejection of the transeunt character of force, Leibniz's impact theory could be worked out consistently only on the assumption of extended particles. The last resort, therefore, of Leibniz's dynamics was ultimately Huygens's mechanism of extended atoms, a theory that is incompatible with the central ideas of Leibniz's monadology. The only way out of this impasse was to reject the idea of extended particles altogether and to conceive forces as associated with mathematical points as their centers, a conception that naturally presupposes action at a distance as the only mode of dynamic interaction. It was Roger Joseph Boscovich who advanced the real Leibnizian theory of dynamics, although ulti-

---

hand, is nothing but the continuation or diffusion of the substance already presupposed, which thrives, withstands, that is, resists, and can consequently never by itself constitute substance." *Specimen dynamicum,* pars I, in *Opera omnia,* vol. 3, p. 315.

[24] Bertrand Russell, *A critical exposition of the philosophy of Leibniz* (Allen and Unwin, London 1937), chap. 7.

mately, as we shall see, Boscovich's concept of force is relational rather than dynamical.

The occasion that led Boscovich to the formulation of his dynamic theory of forces was Father Carolus Scherffer's suggestion of studying the center of oscillation (or center of percussion) of solid bodies. This research led Boscovich to the investigation of the phenomenon of impact or collision of two bodies, a subject dealt with in his *De viribus vivis*.[25] In sections 16 and 17 of his *magnum opus, A theory of natural philosophy,* Boscovich writes:

In the year 1745, I was putting together my dissertation "De Viribus vivis," and had derived everything that they who adhere to the idea of Leibniz, and the greater number of those who measure "living forces" by means of velocity only, derive from these "living forces" . . . I then began to investigate somewhat more carefully that production of velocity which is thought to arise through impulsive action, in which the whole of the velocity is credited with being produced in an instant of time by those, who think, because of that, that the force of percussion is infinitely greater than all forces which merely exercise pressure for single instants. It immediately forced itself upon me that, for percussions of this kind, which really induce a finite velocity in an instant of time, laws of their actions must be obtained different from the rest.

However, when I considered the matter more thoroughly, it struck me that, if we employ a straightforward method of argument, such a mode of action must be withdrawn from Nature, which in every case adheres to one and the same law of forces, and the same mode of action. I came to the conclusion that really immediate impulsive action of one body on another, and immediate percussion, could not be obtained, without the production of a finite velocity taking place in an indivisible instant of time, and this would have to be accomplished without any sudden change or violation of what is called the Law of Continuity; this law indeed I considered as existing in Nature, and this could be shown to be so by a sufficiently valid argument.[26]

[25] P. Rogerius Josephus Boscovich, S.J., *De viribus vivis, dissertatio habita in Collegio Romano Soc. Jesu* (Rome, 1745); reprinted, 1747, in *Commentariis Academiae Bononiensis,* vol. 2, pars 3, secs. 21–49.

[26] *Theoria philosophiae naturalis redacta ad unicam legem virium in natura existentium, auctore P. Rogerio Josepho Boscovich* (editio Veneta prima, 1763). A first edition was published in Vienna in 1758. We follow, however, the Venetian edition, which was revised and enlarged by Boscovich himself and has been translated into English by J. M. Child: *Theoria philosophiae naturalis* (Chicago and London, 1922; Latin-English edition), p. 45.

In order to find out whether the change of velocity in the impact of two bodies changes continuously or discontinuously, Boscovich considers two equal bodies, one with a speed of 6 degrees and the other with a speed of 12 degrees, moving in the same direction on a straight line, the slower in front of the faster. According to the law of conservation of momentum, well established by experiment, both bodies will move on with the common speed of 9 degrees after impact. Let us now study, says Boscovich, how the velocity of the more swiftly moving body changed from 12 degrees to 9. Let us assume for a moment that the fast body behind approaches the other with an undiminished velocity and comes into absolute contact with the former body which was in front. Then clearly its speed must have changed quite abruptly from 12 to 9 degrees without any intermediary degrees.

For it cannot possibly happen that this kind of change is made by intermediate stages in some finite part, however small, of continuous time, whilst the bodies remain in contact. For if at any time the one body then had 7 degrees of velocity, the other would still retain 11 degrees; thus, during the whole time that has passed since the beginning of contact, when the velocities were respectively 12 and 6, until the time at which they are 11 and 7, the second body must be moved with a greater velocity than the first; hence it must traverse a greater distance in space than the other. It follows that the front surface of the second body must have passed beyond the back surface of the first body; and therefore some part of the body that follows behind must be penetrated by some part of the body that goes in front. Now, on account of impenetrability, which all Physicists in all quarters recognize in matter, and which can be easily proved to be rightly attributed to it, this cannot happen.[27]

Our assumption of undiminished velocity right up to impact, declares Boscovich, leads us thus to the conclusion that velocity changes abruptly. This result, however, violates the law of continuity, which denies the passage from one magnitude to another without passing through intermediate stages. In fact, claims Boscovich, it would amount to saying that the body would be bound to have 12 degrees of velocity, and 9, at one and the same

[27] *Ibid.*, pp. 45–47.

time.[28] These considerations show, consequently, that the velocity of the more swiftly moving body must change continuously. Boscovich now demonstrates that this continuous change of velocity cannot take place after the impact, using a similar argument based on the principle of impenetrability, and comes thus to the fundamental result: in every impact of solid bodies a continuous change of velocities takes place before the actual contact! "Oportet, ante contactum ipsum immediatum incipiant mutari velocitates ipsae." [29] But since the cause that changes the state of motion of the body is called force, a force must have been exerted prior to the actual contact. Continuing this analysis Boscovich shows that this force must be mutual and must act in opposite directions, as illustrated by iron and a loadstone which attract each other with the same strength in opposite directions, or by a spring introduced between two balls which exerts an equal action on either. "That universal gravity itself is mutual is proved by the aberrations of Jupiter and of Saturn especially (not to mention anything else); that is to say, the way in which they err from their orbits and approach one another mutually." [30] In this last example the force exerted is attraction, while in the case of the impact of two bodies the force is repulsive, since it imparts to the bodies a natural propensity for mutual recession from each other.

So far we have considered Boscovich's approach to our problem and his ways of reasoning in detail; now we continue our discussion of Boscovich's conception of force in a more summarizing manner.

Since real contact never actually takes place, claims Boscovich, the force of repulsion increases indefinitely as the distance decreases. At greater distances the repulsive force changes its sign, so to say, and becomes attractive, thus accounting for the common phenomena of gravity and gravitation. On the basis of

[28] "In ipso momento ea tempora dirimente debuisset habere et 12, et 9 simul, quod est absurdum." *Ibid.*, p. 70 (sec. 63).

[29] *Ibid.*, p. 76 (sec. 72).

[30] *Ibid.*, p. 77 (sec. 74).

Boscovich's theory of forces, it is the same to assert that contact never takes place or to say that it always takes place, since two bodies have always a dynamic connection that depends on their relative distance alone. Each and every particle of the universe is dynamically related to every other particle, the magnitude and direction of the force involved being a function of the distance concerned. Boscovich's theory thus reduces contact phenomena to actions at a distance and consequently eliminates impact as a fundamental concept of mechanics. His theory thus stands in basic contradiction to the conception of those who still adhere to the Cartesian ideal of physical, or, as they used to call it, mechanical explanation. "They call forces like those I propose," says Boscovich,

> nonmechanical, and reject them, just as they also reject the universal gravitation of Newton, for the alleged reason that they are not mechanical, and overthrow altogether the idea of mechanism which the Newtonian theory had already begun to undermine. Moreover, they also add, by way of a joke in the midst of a serious argument derived from the senses, that a stick would be useful for persuading anyone who denies contact.

But Boscovich knows how to handle this argument. He continues:

> It will be enough just for the present to mention that, when a body approaches close to our organs, my repulsive force (at any rate it is that finally) is bound to excite in the nerves of those organs the motions which, according to the usual idea, are excited by impenetrability and contact; and that thus the same vibrations are sent to the brain, and these are bound to excite the same perception in the mind as would be excited in accordance with the usual idea. Hence, from these sensations, which are also obtained in my Theory of Forces, no argument can be adduced against the theory, which will have even the slightest validity.[31]

Boscovich thus defends this thesis that impenetrability is only a spatial expression for the action of a repulsive force. For at very small distances only repulsive forces are active and their magnitude increases indefinitely as the distances decrease. They

[31] *Ibid.*, p. 109 (sec. 127).

are therefore capable of decelerating and arresting the motion of any body however fast it may move.

Then there never can be any finite force, or velocity, that can make the distance between two points vanish, as is required for compenetration. To do this, an infinite Divine virtue, exercising an infinite force, or creating an infinite velocity, would alone suffice.[32]

Having thus outlined the general character of the force, Boscovich begins to investigate the exact law that governs this force. From what has been said it is clear that the intensity of the repulsive branch increases asymptotically with decreasing distance, whereas for great distances, changing over now to an attractive force, it decreases asymptotically according to the inverse-square law of Newtonian mechanics. When representing graphically the dependence of the intensity of the force upon the distance from the force center, Boscovich adopts the convention that repulsion corresponds to a positive ordinate and attraction to a negative one. The resulting curve, representing a continuous function, must clearly intersect the axis of abscissas at least in one point. However, in order to account for cohesion and fermentation as well as for impenetrability and gravitation and to reduce thereby the three great fundamental principles of Newton (gravitation, cohesion, and fermentation) to one single principle, Boscovich assumes that the curve intersects the abscissa in several points, which he calls limit points (*limites*). "The phenomenon of vapour arising from water, and that of gas produced from fixed bodies lead us to admit two more of these limit-points . . . Effervescences and fermentations of many different kinds, in which the particles go and return with as many different velocities, and now approach towards and now recede from one another, certainly indicate many more of these limit-points and transitions." [33] Boscovich thus arrives at the complete law that governs the intensity and the direction of the fundamental force:

[32] *Ibid.*, p. 267 (sec. 360).
[33] *Ibid.*, p. 83 (sec. 79).

Now the law of forces is of this kind: the forces are repulsive at very small distances, and become indefinitely greater and greater, as the distances are diminished indefinitely, in such a manner that they are capable of destroying any velocity no matter how large it may be, with which one point may approach another, before ever the distance between them vanishes. When the distance between them is increased, they are diminished in such a way that at a certain distance, which is extremely small, the force becomes nothing. Then as the distance is still further increased, the forces are changed to attractive forces; these at first increase, vanish, and become repulsive forces, which in the same way first increase, then diminish, vanish, and become once more attractive; and so on, in turn, for a very great number of distances, which are still very minute: until, finally, when we get to comparatively great distances, they begin to be continually attractive and approximately inversely proportional to the squares of the distances. This holds good as the distances are increased indefinitely to any extent, or at any rate until we get to distances that are far greater than all the distances of the planets and comets.[34]

Figure 5 shows the law graphically. In this figure *A* represents an unextended material point (force center) and *AC* an axis of

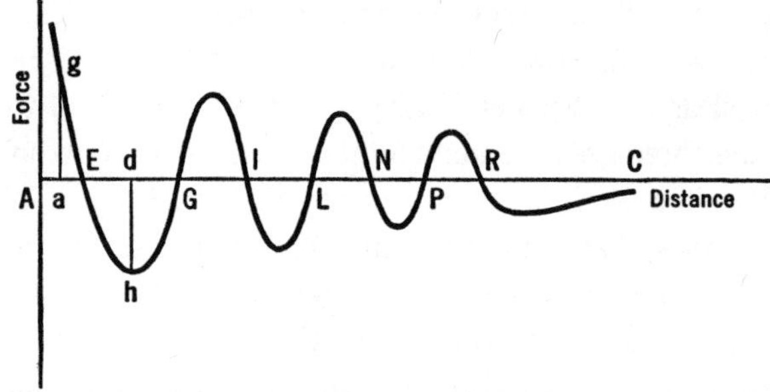

Fig. 5. Boscovich's law of force between particles.

radial distance from it. If another material point is situated at *a*, for example, it will be acted on by a repulsive force of magnitude *ag;* and if it is situated at *d*, it will be attracted toward *A* with a force *dh*. The points *E, G, I,*... are the limit points. Of these *E, I, N,*... are points of stability, as at these points repulsion prevents decrease, and attraction increase, of distance. Two mate-

[34] *Ibid.*, pp. 39–43 (sec. 10).

rial points situated at *A* and *E*, for instance, possess a certain degree of permanency of form and constitute what Boscovich calls a particle of the first order, the first stage in the explanation of the coalescence of groups of points into tenacious masses. Thus the notion of a large system of cohering material points is interpreted as a "little solid mass." [35]

In our discussion of the Boscovichian theory, so far, we have used the term "force" without asking ourselves what actually was Boscovich's definition of this term. It certainly could not be Newton's definition, since Boscovich's material points, although endowed with inertia, still have no mass in the Newtonian sense of the word. Boscovich's definition of force is given in section 9 of his *Theory of natural philosophy* where he says:

> I therefore consider that any two points of matter are subject to a determination to approach one another at some distance, and in an equal degree recede from one another at other distances. This determination I call "force"; in the first case "attractive," in the second case "repulsive"; this term does not denote the mode of action, but the propensity itself, whatever its origin, of which the magnitude changes as the distances change.[36]

Thus for Boscovich "force" is a determination, a propensity of approach or recession, and is measured by the acceleration produced. Nowhere in his writings does Boscovich attempt to explain the cause of these determinations; he merely postulates their existence. Thus, strictly speaking, the ordinates in Fig. 5 on the curve of forces represent accelerations only. That this is indeed a valid interpretation is shown by Boscovich's remark that "the area of a curve, whose abscissae represent distances and

[35] "Quo pacto si multa etiam puncta cohaereant inter se, tuebuntur utique positionem suam, et massam constituent formae tenacissimam, ac eadem prorsus phaenomena exhibentem, quae exhiberent solidae massulae in communi sententia." *Ibid.*, p. 130 (sec. 165).

[36] "Censeo igitur bina quaecunque materiae puncta determinari aeque in aliis distantiis ad mutuum accessum, in aliis ad recessum mutuum, quam ipsam determinationem appello vim, in priore casu attractivam, in posteriore repulsivam, eo nomine non agendi modum, sed ipsam determinationem exprimens, undecunque proveniat, cujus vero magnitudo mutatis distantiis mutetur." *Ibid.*, p. 38 (sec. 9).

ordinates forces, represents the increase or decrease of the square of velocity."[37] In modern mathematical symbolism it means:

$$\text{Area} = \int a\,dx = \int \frac{dv}{dt}\,dx = \int v\,dv \propto v_2^2 - v_1^2.$$

If we, furthermore, recall that Boscovich does not attribute volume[38] or mass to material points, we recognize that his mechanics is consequently pure kinematics, based on the primitive notions of material points and of forces, that is, accelerations, of approach or recession. Ultimately, as we shall see, Boscovich's concept of force is merely relational or functional, as implied by his use of the term "determination" in the definition of force. From the philosophical point of view his theory may be referred to as one of mathematical idealism.

From the standpoint of physics, however, we may claim that a physical theory, based on the notion of force as its most fundamental conception, may be called "dynamic," even if it does not interpret its fundamental concept as a metaphysical entity. Such a criterion of classification and such terminology would be wholly in accordance with our general conception of the nature of science as outlined in the first chapter of this book. Since impenetrability and extension, in Boscovich's view, are merely spatial expressions of forces, "force" is consequently more fundamental than "matter," which, incidentally, at least in the traditional Cartesian sense, has no place at all in his theory. Boscovich's theory of physical sciences may therefore be rightly called "dynamic."

As Boscovich himself stated in the opening passages of his *Theory of natural philosophy,* he proposed to present "a system that is midway between that of Leibniz and that of Newton."[39] Immanuel Kant's treatment of the concept of force was carried

[37] *Ibid.,* p. 105, note f (sec. 118).

[38] "Prima elementa materiae mihi sunt puncta prorsus indivisibilia, et inextensa, quae in immenso vacuo ita dispersa sunt, ut bina quaevis a se invicem distent per aliquod intervallum." *Ibid.,* p. 36 (sec. 7).

[39] *Ibid.,* p. 35 (sec. 1).

out at almost the same time with a similar purpose. In his *Thoughts on the true estimation of living forces*[40] Kant steers a middle course between the Cartesians and the Leibnizians in their dispute about the true measure of force, and in his *Metaphysical foundations of natural science*,[41] written after his precritical period, Kant aims at a philosophical foundation of Newtonian physics, while some of his preconceptions still exhibit clearly a strong influence of the philosophy of Leibniz and Wolff.

The major objective in Kant's *Thoughts on the true estimation of living forces* is of little direct concern for us at present. By an erroneous classification of motions into two kinds, one that persists in the body to which it was communicated and continues indefinitely, and one that ceases with the cessation of the external force that produces it, Kant attempts to do justice to both the Cartesians and the Leibnizians; the Leibnizian measure of mv^2, according to Kant, applies to forces producing motions of the first kind, whereas the Cartesian measure applies to forces producing motions of the second type. Kant accepts the Leibnizian concept of living force as essential to matter and agrees with Leibniz's dictum: "Est aliquid praeter extensionem, imo extensione prius."[42] And, like Boscovich, he comes to the conclusion that "it is easily proved that there would be no space and no extension, if substances had not force whereby they can act outside themselves. For without a force of this kind there is no connection, without this connection no order, and without this order no space."[43] It is important to note that in this work force is for Kant the most fundamental concept and basic for further

[40] Immanuel Kant, *Gedanken von der wahren Schätzung der lebendigen Kräfte und Beurteilung der Beweise, deren sich Leibniz und andere Mechaniker in dieser Streitsache bedient haben* (Königsberg, 1747; not 1746, the date given on the title page; April 22, 1747 is the date of the dedication). Some passages of this work have been translated into English by John Handyside, *Kant's inaugural dissertation and early writings on space* (Chicago and London, 1929).

[41] Immanuel Kant, *Die metaphysischen Anfangsgründe der Naturwissenschaft* (Riga, 1786), translated into English by Ernest Belfort Bax, *Kant's Prolegomena and Metaphysical foundations of natural science* (London, 1909).

[42] G. W. Leibniz, *Specimen dynamicum,* p. 315.

[43] *Kant's inaugural dissertation,* p. 10.

inferences. However, in his critical period the order of these dependences is reversed and the concept of force appears at the end of the chain of inferences.

In the *Metaphysical foundations of natural science* Kant attemps to account for the difference of individual existences by the combinations of two primary forces, attraction and repulsion. In proposition 1 of the second division of this work, dealing with the metaphysical foundations of dynamics, Kant declares: "Matter fills a space, not by its mere existence, but by a special moving force." What he is referring to in this proposition is really impenetrability. In fact, in the demonstration following the proposition, Kant declares: "The resistance offered by a matter in the space which it fills, to all impression of another (matter), is a cause of the motion of the latter in the opposite direction; but the cause of a motion is called moving force." Thus impenetrability is reduced to a moving force and is not conceived as a result of mere existence. Kant, in this context, rejects Johann Heinrich Lambert's contention that the property of matter which makes it fill a space is a solidity assumed to be in everything that exists in the outer world of sense. According to Lambert's notion, says Kant,

> the presence of something real in space must carry with it this resistance by its very conception, in other words according to the principle of contradiction; and must exclude the coexistence of anything else, in the space of its presence. But the principle of contradiction does not preclude any matter from advancing, in order to penetrate into a space in which another (matter) exists. Only when I attribute to that which occupies a space, a power of repelling everything externally movable which approaches it, do I understand how it involves a contradiction, that in the space which a thing occupies, another (thing) of the same kind should penetrate.[44]

The core of the dispute between Lambert and Kant is again the problem of consubstantiality, this time, however, without any reference or implications to theology. For Lambert, the impossibility of two things occupying the same position at the same time

[44] *Kant's Prolegomena,* p. 170.

is a consequence of the law of contradiction, while for Kant it is the result of the repulsive forces associated with either of the contesting parts of matter. The law of contradiction itself, contends Kant, cannot keep back any part of matter from approaching the place that is already occupied by some other part. Hermann Lotze, in his *Metaphysics* comments on Kant's criticism of Lambert: "This objection is not quite fair. We should not expect the physical impossibility referred to to be produced *by* the Principle of Contradiction, but only *in accordance with* that principle and *by* the fact of solidity which, for practical purposes, we assume as an attribute of Real Existence."[45]

In the later sections of his treatise Kant defines the expansive force or elasticity of matter as the force of the extended by virtue of the repulsion of all its parts. Whereas Boscovich assumes the asymptotic character of the repulsive force, Kant declares that by indefinitely increasing compressive forces opposing the forces of extension (expansive forces) matter can be compressed to infinity, but never penetrated. Further he shows that the very existence of matter requires a force of attraction as its second essential fundamental force. Otherwise, owing to the repulsive force, if acting without opposition, matter would dissipate itself to infinity and no assignable quantity of matter would be met with in any assignable space. These are the only two forces that can be conceived, claims Kant. For if two particles regarded as points in space exert upon each other motive forces, these and the motions produced must always be regarded as distributed in the straight line joining these two points. In this straight line, then, only two kinds of motion are possible, approach or recession, corresponding to attractive or repulsive forces. In the subsequent chapters, Kant tries to give an a priori derivation of Newton's three laws of motion and discusses finally the question of transference of motion, that is, the problem of impact. He rejects emphatically the uncritical theory of the so-called transfusionists of motion who argue that motion can be transferred

[45] R. H. Lotze, *Metaphysics* (Oxford, 1884), p. 304.

from one body to another "as water from one glass into the other," [46] a contention that already opposes the scholastic principle: "accidentia non migrant e substantiis in substantias." In its stead Kant suggests an explanation similar to that advanced by Boscovich. Kant's theory of the communication of motion, which "at the same time carries with it as its necessary condition the law of the equality of action and reaction," [47] concludes his discussion of the metaphysical foundations of dynamics.

When comparing Kant's doctrine of forces with Boscovich's theory, we of course note that Kant from the very beginning presupposes two kinds of force, different from each other in their fundamental quality, whereas Boscovich assumes the existence of only one single force. In fact, it was natural to ask, why does the Boscovichian force change over at a certain distance from repulsion to attraction, back again, and so on? No answer could be given to this question. It seemed somehow more comprehensible and less baffling for the imagination to ground this essential difference in the source itself. There were, however, other serious objections that were raised against both these theories with equal weight: it was argued that a dynamic theory such as these could never account for the inertial effects exhibited by atoms and by gross matter. Thus James Clark Maxwell, in his *Theory of heat,* declares: "It is probable that many qualities of bodies might be explained on this supposition, but no arrangement of centres of force, however complicated, could account for the fact that a body requires a certain force to produce in it a certain change of motion, which fact we express by saying that the body has a certain measurable mass." [48] Another important problem, raised in particular with respect to the increasing quantity of experimental information in the area of electrodynamics during the first half of the nineteenth century, was this question: are all forces necessarily central forces? It seemed difficult, on the as-

[46] *Kant's Prolegomena,* p. 228.
[47] *Ibid.,* p. 227.
[48] J. C. Maxwell, *Theory of heat* (London, 1902), p. 86.

sumption of exclusively central forces, to explain permanent deformations, crystallizations, etc., apart from electrodynamic phenomena. Furthermore, the very concept of motion and displacement seemed to be incompatible with the assumption of force centers. If force was the ultimate element of physical reality, it was not nonsensical to ask, as in fact did A. Spir: "Can a force be moved and displaced from one place to another?"[49]

It is therefore not surprising that the theory of dynamism found only a few adherents among serious scientists, and practically none who remained lifelong enthusiastic advocates of its ideas. Even A. Barré de Saint-Venant, who revived and simplified Boscovich's theory in the middle of the last century, abandoned dynamism eventually.[50] Boscovich's doctrine did enjoy a short period of revival in connection with the so-called rariconstant theory of intermolecular action in elasticity, which aimed at an explanation of elastic isotropy by one single material constant. It was in this respect that Saint-Venant adopted Boscovich's ideas and advanced a theory of elasticity based on a Boscovichian curve of force which intersects the axis of distance in only one point.[51] In another paper, to which we have already referred on page 133, Saint-Venant tries to explain crystallization on the basis of Boscovich's theory adding that the acceptance and thorough comprehension of that theory will save science "from the two most serious philosophical aberrations of our time and of ancient times, pantheism and materialism."[52] This remark shows clearly that the conceptions of Newton's theological commentators had not yet lost all power. Since dynamism could easily be interpreted as a spiritual doctrine, Boscovich's theory

[49] Afrikan Spir, *Denken und Wirklichkeit* (Leipzig, 1877), p. 405.

[50] Cf. H. E. Padé, *Revue générale des sciences 15,* 765 (1905).

[51] A. Barré de Saint-Venant, *Mémoire sur la question de savoir s'il existe des masses continues et sur la nature probable des dernières particules des corps* (Paris, 1844).

[52] "Des deux principales et plus funestes aberrations philosophiques de notre temps et des temps anciens, le panthéisme et le matérialisme." Barré de Saint-Venant, "De la constitution des atomes," *Annales de la Société scientifique de Bruxelles 2* (1878), p. 74.

stood in high favor with some clerical writers on physical science, for example, with l'Abbé Moigno.[53] Also the problem of the ultimate constitution of matter led some nineteenth-century theoreticians to a sympathetic attitude toward Boscovich's ideas. Thus Augustin Cauchy, in his *Sept leçons de physique générale*,[54] finding that infinite divisibility of matter leads to contradictions, comes to the conclusion that the ultimate constituents of matter must be unextended and therefore similar to Boscovich's centers of force.

It seems likely that Michael Faraday, at least temporarily, endorsed Boscovich's theory of force centers. In his letters to Richard Taylor[55] and Richard Phillips[56] he admits to having been persuaded to adopt, at least for the time being, Boscovich's point of view and refers to atoms as centers of forces.[57] It is certainly not unreasonable to assume that his conception of lines of force led him to this conclusion. Yet, in a letter to Tyndall, some ten years later, Faraday rejects any such implications: "You are aware (and I hope others will remember) that I give the lines of force only as representations of the magnetic power, and do not profess to say to what physical ideas they may hereafter point, or into what they will resolve themselves."[58] On the other hand, Tyndall clearly interprets Faraday's opinion as favorably disposed toward dynamism when he says in his *Faraday as a discoverer:* "What do we know of the atom from its force? You imagine a nucleus which may be called *a*, and surround it by forces which may be called *m*; to my mind the *a* or nucleus

[53] M. l'Abbé François Napoléon Marie Moigno, *Leçons de mécanique analytique* (Paris, 1868). Cf. also his dissertation *Sur l'essence de la matière* (Paris 1867), published together with *La matière et la force*, his translation of John Tyndall's *Matter and force.*

[54] Augustin Louis Cauchy, *Sept leçons de physique générale* (Paris, 1868).

[55] Michael Faraday, "A speculation touching electric conduction and the nature of matter," *Phil. Mag. 24,* 136 (1844).

[56] Faraday, "Thoughts on ray-vibrations," *Phil. Mag. 28,* 345 (1846).

[57] *Ibid.*, p. 347.

[58] Faraday, "Magnetic remarks," *Phil. Mag.* [4] *9,* 254 (1855). For Faraday's attitude toward the problem of action at a distance see Mary B. Hesse, "Action at a distance in classical physics," *Isis 46,* 337 (1955).

vanishes and the substance consists of the powers *m*. And, indeed, what notion can we form of the nucleus independent of its powers? What thought remains on which to hang the imagination of an *a* independent of the acknowledged forces?"[59] Tyndall's interpretation of Faraday's view states, in the language of physics, what philosophy had claimed long ago.[60] Essentially it is the idea expressed by Schelling so vividly in his *Ideas towards the philosophy of nature,* itself a dynamical explanation of natural phenomena in the spirit of the Kantian doctrine as opposed to the mechanico-atomistic view: "It is a mere delusion of the phantasy that something, we know not what, remains after we have denuded an object of all the predicates belonging to it."[61]

It is this philosophical tenet that induced numerous idealistic philosophers of the nineteenth century in their attitude toward scientific empiricism to embrace a sympathetic view of the dynamism of Boscovich or Kant. Eduard Beneke and Gustav Theodor Fechner may be named as representatives of this school. Fechner's doctrine of atomism, based on the Spinozan-Kantian idea that soul and body are but two differents modes of the appearance of the real, is fundamentally a theory of dynamism, for the atom is not a thing-in-itself, but only the simplest possible phenomenon in "consciousness," having position without extension.[62]

Seven years later, in 1862, the first edition of Herbert Spencer's *First principles* appeared; in it the philosophical exploitation of the concept of force reached its zenith. In cognizance of the two new fundamental ideas of that century, the idea of evolution and

[59] John Tyndall, *Faraday as a discoverer* (London, 1868), p. 123.

[60] See, for example, John Locke, *Essay concerning human understanding* (1690), book II, chaps. XXIII and XXIV.

[61] "Es ist eine blosse Täuschung der Einbildungskraft, dass, nachdem man einem Object die einzigen Prädikate die es hat, hinweggenommen hat, noch Etwas, man weiss nicht 'was, von ihm zurückbleibe." F. W. J. von Schelling, *Ideen zur Philosophie der Natur* (1797); *Sämtliche Werke,* vol. 1 (Stuttgart and Augsburg, 1856), p. 18.

[62] G. T. Fechner, *Ueber die physikalische und philosophische Atomenlehre* (Leipzig, 1855).

the principle of "correlation of forces," as the law of conservation of energy was called at that time, Spencer contends that there must be something at the back of the evolutionary drama which we witness, something that is both a principle of activity and a permanent nexus. We are certain of its existence, but do not know what it is. It is an unknowable force, producing knowable likenesses and differences among its manifestations, constituting subject and object, ego and nonego, mind and matter. The sense of effort, our subjective symbol for objective force in general, is correlative to that ultimate, but unknowable, reality.

We come down then finally to Force, as the ultimate of ultimates. Though Space, Time, Matter, and Motion, are apparently all necessary data of intelligence, yet a psychological analysis shows us that these are either built up of, or abstracted from, experiences of Force. Matter and Motion, as we know them, are differently conditioned manifestations of Force.[63]

The ultimate element in our idea of body as distinguished from space, according to Spencer, is the power of neutralizing that which we know as our own muscular strain. Resistance to our effort is the fundamental property of matter,

a resistance we are to symbolize as the equivalent of the muscular force it opposes. In imagining a unit of matter we may not ignore this symbol, by which alone a unit of matter can be figured in thought as an existence. It is not allowable to speak as though there remained a conception of an existence when that conception has been eviscerated — deprived of the element of thought by which it is distinguished from empty space. Divest the conceived unit of matter of the objective correlate to our subjective sense of effort, and the entire fabric of physical conceptions disappears.[64]

We have already mentioned (page 10) Spencer's extreme vagueness and almost unparalleled laxity in the use of the term force. His promiscuous use of the term force both for the notion of force in the Newtonian sense and for energy leads him to serious inconsistencies and his reader to unfortunate miscompre-

[63] Herbert Spencer, *First principles* (fourth ed., New York, 1895), p. 169.
[64] *Ibid.*, p. 192.

hensions. "While recognizing this fundamental distinction between that intrinsic force by which body manifests itself as occupying space, and that extrinsic force distinguished as energy; I here treat of them together as being alike persistent."[65] This indistinctness in word and notion, as far as the exact sciences are concerned, as well as the fact that his philosophy bears more upon biology, psychology, and sociology, made his doctrine of little appeal to the professional physicist. The same can be safely said with respect to other theories of physical dynamism as expounded within the framework of late nineteenth-century philosophy, and in particular the neocriticism of Renouvier or Magy.[66] The conception of force as the primordial element of physical reality, advanced by Leibniz, Boscovich, Kant and their followers, was not very fruitful and productive for the advancement of theoretical physics. It was a construct that was not easily assimilable into a conceptual scheme of operational import, and apparently remains such even in our atomic age which succeeded in releasing "the force of the atom." As far as physics proper is concerned, every modern physicist, most certainly, would agree with Thomson and Tait in calling such dynamic doctrine an "untenable theory."[67]

[65] *Ibid.*

[66] C. B. Renouvier, "Les principes de la nature," in *Essais de critique générale* (1854–1864; Paris, 1912). F. Magy, *De la science et de la nature* (Paris, 1865). Cf. J. G. F. Ravaisson-Mollien, *Rapport sur la philosophie en France au dix-neuvième siècle* (Paris, 1868), chapter 20.

[67] Lord Kelvin and P. G. Tait, *Treatise on natural philosophy* (Cambridge University Press, Cambridge, 1923), part II, p. 214.

CHAPTER 10

MECHANISTIC THEORIES OF FORCE (GRAVITATION)

In our exposition of post-Newtonian concepts of force and gravitation we have discussed so far two major trends of thought, both of which originated, or at least found their scientific support, in Newton's foundation of classical mechanics: the religious school of commentators who interpreted Newton's concept as a vindication of their natural theology, and the school of dynamism which conceived force as the ultimate essence of physical reality. We shall discuss a third school, one that interprets force as a pure relational concept in Chapter 11. In the present chapter we examine the development of mechanistic theories of gravitation.

From the time of Robert Hooke (who, strictly speaking, advanced his mechanistic theory before the publication of Newton's *Principia*) to the present day,[1] innumerable mechanistic theories of gravitation have been published, each trying to succeed where Newton failed. The authors of these treatises oppose the theory of action at a distance, conceiving forces of push and pull as more fundamental and basic for their explanation of the transmission of force. In the majority of these theories action at a distance is not claimed to be self-contradictory on the ground

[1] The present author received only two years ago, while lecturing at Harvard University, a two-volume mimeographed "Kinetic theory of gravitation" for review purposes.

that spatio-temporal separation is derived from causal separation; in other words, their rejection of action at a distance is not preceded by a logical analysis of this concept with respect to causality. Therefore these theories may be spoken of as still precritical, so to say, this being, of course, the reason why we postponed the discussion of the relational conception of force to the following chapter.

Needless to say, it cannot be the purpose of the present chapter to give an exhaustive and detailed description of all mechanistic theories of gravitation advanced during the last two hundred years. We shall confine our discussion to the most important and most interesting theories of this kind, showing how much intellectual ingenuity and industry has been spent in the endeavor to reduce action at a distance to contiguity.

Robert Hooke's theory of gravitation, published posthumously, may be taken as the first instance of this long list of kinetic theories. In analogy to his belief that small bodies floating on the surface of agitated water tend toward the center of the disturbance, Hooke postulated that all parts of the earth execute small, imperceptible, and rapid vibrations or spherical pulsations with the center of the earth as their center. The ether contained in the terrestrial matter as well as the ether surrounding the earth vibrates with amplitudes that decrease with increasing distance from the center of vibration. As a result of these vibrations, not only parts of the earth itself but also all tangible matter above the surface of the earth is moved toward the center of the vibrations, as small floating bodies are pushed toward the center of water waves. Since the area of these spherical vibrations increases with the square of their radius, gravity conforms to the inverse-square law of distance from the center of the earth. "By which radiating vibration of this exceeding fluid and yet exceeding dense matter, not only all the parts of the earth are carried or forced down toward the centre . . . From hence I conceive the power thereof to be always reciprocal to the area or superficies of the orb of propagation, that is duplicate of the distance."[2] Hooke asserts the proportionality of gravitational force with the quantity of matter of the "attracted" body, since the ether penetrates it completely and the vibrations affect all of his parts.[3] His theory is obviously incomplete, since it does not explain the salient point, namely, exactly how these vibrations cause centripetal motion; yet it was an audacious attempt to explain, not only gravitation, but also solidity and hardness and other qualities of matter,[4] by the concept of motion.

The publication of Newton's *Principia* did not put a sudden end to the acceptance of the Cartesian physics of ethereal vortices. Newton's derivation of Kepler's laws brought in its train a great number of publications, mostly by French physicists who tried to derive Kepler's laws on the assumption of vortical motion. Naturally, these papers also contained various speculations about the nature of gravitation. Thus Philippe Villemot, in his astronomical treatise *Nouveau système*,[5] reduces gravitation to a difference of pressure of the fluid of which the vortex is constituted. His explanation is similar to that advanced by Newton in his letter to Boyle, quoted on page 135.

In 1734 the French Academy of Sciences awarded a prize to John Bernoulli's memoir, "The mutual inclination of the planetary orbits,"[6] the first three parts of which attempt to explain the cause of gravitation while the fourth part deals with the

[2] Robert Hooke, *Posthumous works* (London, 1705), pp. 184–185.

[3] "So the momentum of every body becomes proportionated to its bulk or density of parts." *Ibid.*

[4] Cf. Hooke, *An attempt for the explication of the phenomena observable in an experiment published by the Hon. R. Boyle* (London, 1661; Latin trans. by Behem, Amsterdam, 1662); *Micrographia or some physiological descriptions of minute bodies* (London, 1665); *Lectures de potentia restitutiva, or of spring, explaining the power of springing bodies* (London, 1678).

[5] Philippe Villemot, *Nouveau système, ou nouvelle explication du mouvement des planètes* (Lyons, 1707).

[6] John Bernoulli, "Essai d'une nouvelle physique céleste servant à expliquer les principaux phenomènes du ciel, et en particulier la cause physique de l'inclination des orbites des planetes par raport au plan de l'équateur du soleil," in G. Cramer, ed., *Johannis Bernoullii opera omnia* (Lausanne and Geneva, 1742), vol. 3, p. 263.

astronomical problem proposed. Rejecting both the Cartesian and the Newtonian concepts of gravitation, Bernoulli explained gravitation from the impulsions of a central torrent, a continual descent from the heavens of "a copious and impetuous rain of pellets, driven inward by the shocks of molecules from surrounding vortices." At the boundaries of the solar vortex, centripetal motion is imparted to these small pellets, constituting the central torrent; they penetrate the pores of gross matter to varying depths, before being therein arrested, and cause the drift of material bodies in the direction toward the center of the earth. It would lead us too far afield to show how this conception accounts for the quantitative aspects of gravitation in detail. The mechanism is complicated and certainly does not conform to Bernoulli's emphatic statement, written in the same work: "In physical science one should banish the practice of explaining phenomena by chimerical principles more obscure than those presented for investigation." [7]

With the gradual recognition that Cartesian physics is incapable of accounting for Kepler's planetary laws and with the general acceptance of Newton's theory of gravitation, the number of publications based on vortical mechanics decreases rapidly. Although Henry Cavendish's classical experiment of "weighing the earth," that is, the determination of the constant of universal gravitation, was performed only toward the end of the eighteenth century, the idea of the experiment is much older. Thus it is known that the construction of an apparatus for the performance of this experiment was envisioned in 1768 by John Michell. The experimental verification that two material spheres of ordinary size attract each other in accordance with Newton's law of gravitation gave the *coup de grâce* to all vortical explanations of gravitation, since the assumption of vortices associated with each and every material particle individually led to insurmountable philosophical difficulties, if not absurdities. This, however, does not

[7] Bernoulli, *Opera omnia,* vol. 3, sect. XXXII, p. 288.

imply that the search for mechanistic explanations was given up. It only meant the abandonment of ethereal vortices of planetary size as basic for the explanation of gravitational phenomena. Instead of Bernoulli's "central torrent" of radially converging ethereal particles, a homogeneous torrent of ethereal corpuscles, traversing space in straight lines, was still a possibility to be tried out.

A mechanistic theory of gravitation of this kind was advanced in 1747 by Georges Louis Lesage. In this ingenious theory an infinite number of "ultramundane particles" of transcendent minuteness and exceeding speed was assumed to traverse space in all directions in straight lines. A single body placed far away from all other tangible matter would remain at rest, since it would be equally affected by these particles in all directions. As soon, however, as a second body is placed nearby, it will screen off the particles and cast, so to say, a mechanical shadow on the former body. This latter will therefore be impelled only by the impinging particles from the other side and hence will approach the screening body with a force that varies inversely with the square of the distance. Moreover, it is obvious that this effect is reciprocal, that is to say, both bodies through their mutual screening effect will "gravitate" toward each other, thus accounting for universal gravitation. In a letter dated January 15, 1747, Lesage reported to his father: "Eureka! Eureka! Never was I so satisfied as at this moment when I succeeded in explaining the principle of universal gravitation by the simple laws of rectilinear motion."

Lesage published his kinetic explanation two years later in an essay *Sur l'origine des forces mortes*[8] and later in a paper entitled "Loi qui comprend toutes les attractions et répulsions."[9] A detailed exposition, reviving the atomic theory of Democritus and Epicurus to account for all physical forces and processes, was

[8] Georges Louis Lesage, "Essai sur l'origine des forces mortes" (1749; manuscript, Library of Geneva).

[9] Published in the *Journal des Sçavants* (April 1764).

published by him in his famous "Lucrèce Newtonien" [10] in 1782, in which he says:

I propose to show that if the first Epicureans had only as sound ideas concerning cosmography as some of their contemporaries to whom they neglected to pay attention, and if they had known only a portion of the facts of geometry that were already of common knowledge, they would, very probably, have discovered without effort the laws of universal gravitation and its mechanical cause. Laws, of which the formulation and demonstration are the greatest glory of the mightiest genius that ever existed; and cause which, after having long constituted the ambition of the greatest physicists, is today the despair of their successors.

Yet, brilliant and simple as his ideas looked at first glance, later thought showed their insufficiency. First of all, simple screening effects could not account for the gravitational phenomena encountered if three bodies were placed in one straight line, as happens, for instance, in an eclipse of the moon. Second, the doctrine of what Lesage called "gravific fluid" could not account for the mass dependence of gravitational force. Much labor has been spent in revising and amending Lesage's theory. For example, solid matter was supposed to be porous to account for the three-body problem; absorption effects were incorporated to explain the mass dependence; and so on. When it was still assumed that the theory could be worked out to meet all requirements, it was hailed as the greatest achievement. Thus, Pierre Prevost, in his biography of Lesage, published two years after Lesage's death,[11] proclaimed: "I pause at the foot of this majestic edifice with a sentiment of hope; persuaded that the labors of the founder will not be suffered to perish, and that men of genius will share with me the admiration it has inspired." Fifty years later, when the insufficiency of this theory was made evident beyond any doubt, Sir John Herschel wrote:

The hypothesis of Lesage which assumes that every point of space is penetrated at every instant of time by material particles *sui generis,*

[10] Published in *Mémoires de l'Académie royale des Sciences et Belles Lettres de Berlin, pour 1782* (1784).

[11] Pierre Prevost, *Notice de la vie et des écrits de G. L. Le Sage* (*suivie d'un opuscule de Le Sage sur les causes finales*) (Geneva, 1805).

moving in right lines in every possible direction, and impinging upon the material atoms of bodies, as a mode of accounting for gravitation, is too grotesque to need serious consideration; and besides will render no account of the phenomena of elasticity.[12]

Characteristic of a serious mathematician is Leonhard Euler's reserved attitude toward the problem of gravitation. In his numerous writings on problems of mechanics, the question of gravitation is mentioned explicitly only in two of his works, his *Dissertatio de magnete*[13] and his *Letters to a German Princess.*[14]

In his dissertation on the magnet Euler reverts to an ether theory that attempts to explain the various physical agencies, and among them gravitation, by the presence of ether. Although he admits the difficulty of the problem and speaks in very cautious terms, he insists upon the necessity of finding a mechanical cause for gravitation. Similarly he states in his *Letters:*

Let us suppose that before the creation of the world God had created only two bodies, at a distance from each other; that absolutely nothing existed outside of them, and that they were in a state of rest; would it be possible for the one to approach the other, or for them to have a propensity to approach? How could the one feel the other at a distance? Whence could arise the desire of approaching? These are perplexing questions. But if you suppose that the intermediate space is filled with a subtile matter, we can comprehend at once that this matter may act upon the bodies by impelling them. The effect would be the same as if they possessed a power of mutual attraction. Now, as we know that the whole space which separates the heavenly bodies is filled with a subtile matter called aether, it seems more reasonable to ascribe the mutual attraction of bodies to an action which the aether exercises upon them, though its manner of acting may be unknown to us, than to have recourse to an unintelligible property.[15]

[12] Published in *Fortnightly Review* (July 1, 1865), p. 438. Contrary to Herschel's anticipation, Lesage's hypothesis was subsequently elaborated and revised particularly by C. Isenkrahe, *Das Raetsel der Schwerkraft* (Vieweg, Braunschweig, 1879), by S. T. Preston, *Phil. Mag.* [5] *4* (1877), *11* (1894), and by A. Rysánek, *Rept. Exp.-Phys. 24,* 90–115 (1887).

[13] Leonhard Euler, *Dissertatio de magnete* (1743), in *Opuscula varii argumenti* (Berlin, 1746–1751), vol. 3 (1751). In 1744 this dissertation received the prize of the Paris Academy.

[14] Leonhard Euler, *Lettres à une Princesse d'Allemagne, sur divers sujets de physique et de philosophie* (Paris, 1842).

[15] *Ibid.,* letter 68, October 18, 1760.

Euler adopted this cautious and reserved attitude, bordering almost on resignation, because he realized that so far no purely mechanical explanation of gravitation had been afforded. In the search for an explanation by possible other agencies than purely mechanical ones, John Herapath's thermogenetic theory of gravitation is most remarkable. In two essays published in Thomson's *Annals of Philosophy*,[16] Herapath presents a clear exposition of the mechanical theory of heat and reduces gravitation to the effects of temperature differences. A more detailed systematic exposition of his ideas was published by him in his *Mathematical physics*[17] in 1847.

If gravitation, he claims, depends upon the action of an elastic medium, as conceived already by Newton, and if this medium is supposed to have decreasing density the nearer it is situated to the dense bodies of the sun or the planets, there ought to be some reason for this variation of density.

> If this medium be of the same nature as our atmosphere and other gaseous bodies, that is, if it be capable of being expanded by heat and contracted by cold, then the sun being a very hot body, and the heat being so much the greater the nearer we are to him, the density of the medium ought therefore to decrease with a decreasing and increase with an increasing distance, the same as Newton would have it. And because we find by experience that dense solid bodies receive heat more strongly than much rarer ones, particularly than gases, the dense bodies of the planets being heated by the solar rays as well as by the medium, about them, ought it appeared to me, to be hotter than this medium, and consequently ought to produce the same effects on the medium as the sun, though not in so great a degree. Therefore if as Newton imagines, the particles of the planets be impelled toward the sun by the inequality of the pressure on their further and nearer sides, the denser parts of the medium pressing more forcibly than the rarer, the same reason will likewise hold good why bodies should be impelled toward the planets and other material parts of the system.[18]

[16] John Herapath, "On the physical properties of gases," *Annals of Philosophy 8* (1816); "A mathematical inquiry into the causes, laws, and principal phenomena of heat, gases, gravitation," *Annals of Philosophy 17* (1821).

[17] John Herapath, *Mathematical physics; or the mathematical principles of natural philosophy* (London, 1847).

[18] Herapath, "A mathematical inquiry," p. 276.

Needless to say, very elementary mathematical considerations show that Herapath's thermogenetic theory of gravitation is full of deficiencies and inconsistencies.

When one reads the scientific literature in the journals and treatises of the middle of the nineteenth century, the mechanistic explanation of gravitation seems to have been the most engaging and promising problem of the generation. Yet today these papers on the whole have little importance. We shall not discuss the details of these theories, interesting as they often are, but shall refer the reader to the publications of Guyot,[19] Seguin,[20] Boucheporn,[21] Lamé,[22] Waterston,[23] Challis,[24] Glennie,[25] Keller,[26] Croll,[27] Leray,[28] Boisbaudran,[29] and Guthrie.[30] Their common fea-

[19] Jules Guyot, *Éléments de physique générale* (G. Baillière, Paris, 1832).

[20] Marc Seguin, "Considérations sur la tendence qu'éprouvent les molécules matérielles à se réunir entre elles . . . et sur les moyens d'expliquer ces faits par les seules lois de l'attraction newtonienne," *Comptes rendus 27,* 314 (1848); "Considérations sur la loi qui maintient les molécules matérielles à distance," *Comptes rendus 28,* 97 (1849); "Considérations sur l'action qu'exercent des molécules en mouvement sur des molécules qui s'approchent ou s'éloignent l'une de l'autre," *Comptes rendus 29,* 425 (1849); "Considérations sur la détermination des conditions dans lesquelles devraient se trouver les molécules matérielles qui constituent le globe terrestre, pour que les effets de la cohésion des corps cristallisés qui existent à sa surface pussent être expliqués par les lois de l'attraction newtonienne," *Comptes rendus 34,* 85 (1852); "Mémoire sur la cohésion," *Comptes rendus 37,* 703 (1853).

[21] M. F. Boucheporn, "Recherches sur les lois physiques considérées comme conséquences des seules propriétés essentielles à la matière, l'impénétrabilité et l'inertie," *Comptes rendus 29,* 107 (1849); cf. also his *Principe générale de la philosophie naturelle* (Paris, 1853).

[22] Gabriel Lamé, *Leçons sur la théorie mathématique de l'élasticité des corps solides* (Paris, 1852).

[23] J. J. Waterston, "On the integral of gravitation," *Phil. Mag. 15,* 329 (1858).

[24] James Challis, "A mathematical theory of attractive forces," *Phil. Mag. 18,* 321 (1859); "A theory of molecular forces," *Phil. Mag. 19,* 88 (1860); "On the principle of theoretical physics," *Phil. Mag. 23,* 313 (1862). Cf. also his *Principles of mathematics and physics* (Cambridge, 1869).

[25] J. S. Stuart Glennie, "On the principles of the science of motion," *Phil. Mag. 21,* 41 (1861); "On the principles of energetics," *Phil. Mag. 21,* 274, 350 (1861), *22,* 62 (1861).

[26] F. A. E. and E. Keller, "Mémoire sur la cause de la pesanteur et des effets attribués à l'attraction universelle," *Comptes Rendus 56,* 530 (1863).

[27] James Croll, "On certain hypothetical elements in the theory of gravitation," *Phil. Mag. 34,* 449 (1867).

[28] P. Leray, "Théorie nouvelle de la gravitation," *Comptes Rendus 69,* 616 (1869).

ture, almost without exception, is the introduction of hypotheses *ad hoc,* whether as to the structure of matter and the ether, or as to complicated undulatory motions of the ether particles. These theories, moreover, apart from being explanations of *ignotum per ignotius,* and violating the principle of simplicity, propose many conceptions that from the beginning defy all experimental verification because they are devoid of epistemic correlations. And in those few cases where reference is made to some experimental verifiability, the theory proves to be fundamentally wrong. Thus William Crookes thought that the mystery of gravitation could be solved by his famous experiments on radiation, in which he constructed an apparatus called today a radiometer. "It is not unlikely," he says,

> that in the experiments here recorded may be found the key of some as yet unsolved problems in celestial mechanics. In the sun's radiation passing through the quasi-vacuum of space we have the radial repulsive force possessing successive propagation to account for the changes of form in the lighter matter of comets and nebulae . . . but until we measure the force exactly, we shall be unable to say how much influence it may have in keeping the heavenly bodies at their respective distances. So far as repulsion is concerned, we may argue from small things to great, from pieces of pith up to heavenly bodies . . . Although the force of which I have spoken is clearly not gravity solely, as we know it, it is attraction developed from chemical activity, and connecting that greatest and most mysterious of all natural forces, action at a distance, with the more intelligible acts of matter. In the radiant molecular energy of solar masses may at last be found that "agent acting constantly according to certain laws," which Newton held to be the cause of gravity.[31]

Misinterpreting the effect of differential heat absorption by the two sides of the radiometer vanes, Crookes thought that this impulsion of radiation could account for gravitational phenomena. It is, however, interesting to note that Crookes, owing to his faulty interpretation of experimental data, anticipated Lebedew's discovery in 1901 of radiation pressure, which was

[29] M. Lecoq de Boisbaudran, "Note sur la théorie de la pesanteur," *Comptes Rendus 69,* 703 (1869).

[30] Frederick Guthrie, "On approach caused by vibration," *Phil. Mag. 40,* 345 (1870).

[31] William Crookes, *Trans. Roy. Soc.* (*London*) *164,* 527 (1874).

thoroughly investigated by Nichols and Hull in 1903 and more recently by Gerlach and his collaborators in 1923. Furthermore, in an essay on "The mechanical action of light," Crookes already associates his radiative impulsion with cosmological phenomena.

> It may be said that a force like this must alter our ideas of gravitation; but it must be remembered that we only know the force of gravity between bodies such as they actually exist, and we do not know what this force would be if the temperatures of the gravitating masses were to undergo a change. If the sun is gradually cooling, possibly its attractive force is increasing, but the rate will be so slow that it will probably not be detected by our present means of research.[32]

Our survey of mechanistic theories was limited to more or less serious, although clearly not unimpeachable, attempts of explaining gravitation. Apart from these, an enormous number of less respectable hypotheses was promulgated by amateurs and charlatans in semiscientific journals and popular books. Ranging from puerile explications to literary masterpieces, this science fiction of the nineteenth century is not without a charm of its own. We shall mention only one example, for the sake of illustration, the theory of substantialism propagated and propagandized by its author, Wilford Hall. In the view of this writer, ether as the medium of the transmission of gravitational force has to be discarded — not a bad idea — and the force of gravity or gravitation has to be conceived as a substance, though invisible. This theory of "substantialization of gravitational force" is, of course, only one instance of the doctrine of substantialism, according to which all phenomena are to be reduced to substantial realities. Hall disseminated these ideas in a magazine, *The scientific arena* (1886–1888) which he edited, as well as in the journal, *The microcosm*" (1881–1892), of which he was editor in chief (and most prolific contributor). Hall's theory of gravitation is also expounded in his book *The problem of human life,* from which we quote the following trochaic tetrameters of a Hiawatha lilt, forming the core of his theory of gravitation:

[32] William Crookes, "The mechanical action of light," *Quart. J. Sci. 6,* 254 (1876).

Strange that such man as Newton
When conceiving some connection
Linking attrahents together
To account for drawing motion
Could not think just one step further,
Or conceive that gravitation
Might itself be real substance
Of invisible formation—
Chords of force connecting bodies,
Spun from each corporeal atom,
While their molecules, like bobbins,
Reel incessantly these force-threads,
Till the objects thus united
Should be fully brought in contact.[33]

[33] A. Wilford Hall, *The problem of human life* (New York, second ed., 1877), p. 68.

CHAPTER 11

MODERN CRITICISM OF THE CONCEPT OF FORCE

As we have already mentioned, the third school of thought in the interpretation of the concept of force, namely, those who conceive force either as a primitive and irreducible notion or as a purely relational concept, devoid of a separate ontological status and to be defined only operationally, can also be traced back to Newton and his immediate commentators. It is an empirical school of thought, for it emphasizes the importance of experience in contrast to rationalistic argumentation. In fact, Cotes's preface to the *Principia,* which contributed so much to the advancement of the concept of an action at a distance, already contained arguments in favor of the relational concept. The discussion is, of course, about the nature of gravitation; but let us not forget that the controversy about gravitation had an effect upon the development of physics and its philosophical interpretation similar to the effect of the problem of Euclid's fifth postulate, the so-called axiom of parallels, upon the logical development of geometry. Not only was gravitation the most important force to be considered, and for many theoreticians ultimately the only force, but conclusions reached about the nature and status of gravitation had repercussions upon the conception of physical science as a whole, as we shall see in the present chapter.

In defense of Newton's method of treating gravitational force, Cotes, in the preface to the second edition of the *Principia,* says:

But shall gravity be therefore called an occult cause, and thrown out of philosophy, because the cause of gravity is occult and not yet discovered? Those who affirm this, should be careful not to fall into an absurdity that may overturn the foundations of all philosophy. For causes usually proceed in a continued chain from those that are more compounded to those that are more simple; when we are arrived at the most simple cause we can go on no farther. Therefore no mechanical account or explanation of the most simple cause is to be expected or given; for if it could be given, the cause were not the most simple. These most simple causes will you then call occult, and reject them? Then you must reject those that immediately depend upon them, and those which depend upon these last, till philosophy is quite cleared and disencumbered of all causes.[1]

No doubt, Cotes, in this context at least, conceives gravity as a primitive concept in the theoretical system of physical science.

Even before Cotes who wrote his preface in 1713, John Keill, Savilian professor of astronomy in the university of Oxford, one of the first to promulgate Newtonian physics, wrote the following statements in his *Introduction to natural philosophy:*

We shall not be ashamed to use, with the Peripateticks, the terms quality, faculty, attraction, and the like; not that by these words we pretend to define the true and physical cause and modus of action, but as these actions may be augmented and diminished, and therefore since they have the properties of qualities, the same name may not unfitly be applied to them, so that we thereby only mean to express the ratios of the forces or their augmentation and diminution. For example, we may say that gravity is a quality, whereby all bodies are carried downwards, whether its cause arises from the virtue of the central body, or is innate to matter itself; or whether it proceeds from the action of the aether agitated by a centrifugal force, and so tending upwards; or lastly, whether it is produced after any other manner whatever . . . Certainly by the same right, as in algebraical equations we denote the unknown quantities by the letter x or y; and not by a very unlike method we may investigate the intensions and remissions of these qualities, which follow from some certain supposed conditions.[2]

[1] *Sir Isaac Newton's mathematical principles of natural philosophy,* trans. Andrew Motte, rev. Florian Cajori (University of California Press, Berkeley, 1934), p. xxvii.

[2] John Keill, *An introduction to natural philosophy: or philosophical lectures read in the University of Oxford, anno dom. 1700* (London, second ed., 1726), p. 4. The Latin original (*Introductio ad veram physicam*) was published in 1702.

As an illustration of these ideas Keill demonstrates the theorem of the inverse-square law concerning the intensity of any "quality or virtue that is propagated every way in right lines from a center," notwithstanding our complete ignorance of the nature of the quality involved, and "how much soever the modus of operation is concealed from us."[3] Keill's comparison of physical forces with unknowns in algebraic equations (the Latin text says *quantitates incognitae*) is most fortunate, since it shows clearly what Keill meant to say: we can handle, measure, or compute their quantitative aspects without knowing what they really are. They are important only in so far as they express relational or functional dependences of other data in space and time. Even Samuel Clarke, in his controversy with Leibniz, is finally forced to adopt a similar attitude. The issue of the dispute was ultimately the question whether empirical scientific conceptions could serve as a basis for rationalistic metaphysics. When Leibniz attacked Newton's conception of attraction, it was not the principle of gravitation as a scientific construct that was under fire, but gravitational force as a metaphysical entity. In defense of the empirical point of view Clarke replied:

> That the sun attracts the earth, through the intermediate void space; that is, that the earth and sun gravitate towards each other, or tend (whatever be the cause of that tendency) towards each other, with a force which is in a direct proportion of their masses, or magnitudes and densities together, and in an inverse duplicate proportion of their distances; and that the space betwixt them is void, that is, hath nothing in it which sensibly resists the motion of bodies passing transversly through: all this, is nothing but a phenomenon, or actual matter of fact, found by experience. That this phenomenon is not produced *sans moyen*, that is, without some cause capable of producing such an effect; is undoubtedly true. Philosophers therefore may search after and discover that cause, if they can; be it mechanical, or not mechanical. But if they cannot discover the cause; is therefore the effect itself, the phenomenon, or the matter of fact discovered by experience (which is all that is meant by the words attraction and gravitation,) ever the less true?[4]

[3] *Ibid.*, p. 5.

[4] *A collection of papers which passed between the late learned Mr. Leibnitz and Dr. Clarke* (London, 1717), pp. 367–368.

These statements do not deny the possibility of ultimately reducing gravitation to other causes. Yet, Clarke, like his master Newton, here comes very near to the operational concept of force and gravitation, since the importance of the concepts concerned for physics as such is independent of whether they are or are not reduced to further causes.

It was soon recognized that the issue involved transcends in its importance the technicalities of mechanics and physics and has a bearing upon the evaluation of scientific knowledge as a whole. The interpretation of "force," "matter," and "motion" became a problem of general philosophic interest. One of the first to investigate these questions from the more general point of view was George Berkeley, who in his *De motu* (1721) and *Siris* (1744) deals with the philosophy of physical science. Although a great admirer of Newton, Berkeley criticizes Newton's fundamental concepts in a way that sounds most modern. In fact, K. R. Popper, in an article in the *British Journal for the Philosophy of Science*,[5] compares Berkeley's criticism with that of Mach and points to the close similarity of the two standpoints. If we remember the great influence that Berkeley's ideas had upon David Hume, and through him upon the whole of modern philosophy, it is not without interest to note that Berkeley's preoccupation with the problems of the philosophy of science, and with the concepts of force, space, and time in particular, was an important factor in this issue. The criticism of the fundamental concepts of Newtonian dynamics was a subject that engaged his mind from the early age of twenty until his most mature publication, *Siris*, at the age of sixty.[6]

In his immediate objective to uncover the illegitimacy in the use of certain general abstract notions in science, Berkeley di-

[5] K. R. Popper, "A note on Berkeley as precursor of Mach," *British Journal for the Philosophy of Science* 4, 26 (1953/54).

[6] The first edition appeared in April 1744, in London; the title *Siris* ("Chain") was prefixed to the second edition. The original title was: *A chain of philosophical reflexions and inquiries concerning the virtues of tar-water and divers other subjects connected together and arising one from another.*

rected his offensive against the fundamental notions of Newton's dynamics. Berkeley's objections to Newton's conceptions of absolute space[7] and time are of only secondary interest for our present subject. But of greatest importance for us is his concise statement about the use of the concept of force, made in *De motu:* "Force, gravity, attraction and similar terms are convenient for purposes of reasoning and for computations of motion and of moving bodies, but not for the understanding of the nature of motion itself." [8] Berkeley admits that Newton himself adopted this point of view as far as attraction is concerned. He says: "With respect to attraction it is clear that this was not introduced by Newton as a true physical quality, but only as a mathematical hypothesis." [9] But the same holds, Berkeley declares, also for the other terms, and in particular for the word "force." "Force" in his view has the same status in science as the notion of "epicycles" in astronomy; although the use of these terms may lead to correct results, it should not be supposed that they are part of nature itself. "Those who suppose epicycles, and by them explain the motions and appearances of the planets, may not therefore be thought to have discovered principles true in fact and nature." Such inference would be wholly unwarranted, since we may infer a conclusion from our premises, but it will not follow that we can "argue reciprocally, and from the conclusions infer the premises." [10] Physics, or, as Berkeley calls it, "mechanical philosophy," can consequently not supply any causal explanations. "Real efficient causes of the motion . . . of bodies do not in any way belong to the field of mechanics or of

[7] See, for example, G. J. Whitrow, "Berkeley's philosophy of motion," *British Journal for the Philosophy of Science 4*, 37 (1954), or Max Jammer, *Concepts of space* (Harvard University Press, Cambridge, 1954), pp. 106ff.

[8] "Vis, gravitas, attractio, et hujusmodi voces, utiles sunt ad ratiocinia et computationes de motu et corporibus motis; sed non ad intelligendam simplicem ipsius motus naturam." George Berkeley, *De motu,* sec. 17; see *Works of George Berkeley,* ed. A. C. Fraser (Clarendon Press, Oxford, 1901), vol. 1, p. 506.

[9] "Non tanquam qualitatem veram et physicam, sed solummodo ut hypothesi mathematicam." *Ibid.*

[10] George Berkeley, *Siris,* sect. 228; *Works,* vol. 3, p. 230.

experimental science; nor can they throw any light on these." [11] The only objective of physical science, according to Berkeley, is to find out the regularities and uniformities of natural phenomena and to account for the particular appearances "by reducing them under, and shewing their conformity to, such general rules." [12]

All that natural science can supply is an account of the relations among symbols or signs; but the sign should not be confounded with the *vera causa,* the real cause of the phenomena. For real causes are active causes, they are productive, they make things occur. And real causes, as we have seen, are not the subject of physical science. The law of attraction, therefore, is to be regarded as a law of motion, and this only as a rule or method "observed in the productions of natural effects, the efficient and final causes whereof are not of mechanical consideration." [13]

When science describes the motions given in direct experience and states them as simply as possible in rules or formulas, it may be useful to refer to "force" or "action" as an element in such formulas or equations, but then "we are not able to distinguish the action of a body from its motion." [14] Science has no right to assume that bodies possess or exert real forces which enable them to act on each other. A close analysis of what we mean by *body* shows us that it contains nothing of this kind. If bodies are divested of all their qualities, as extension, impenetrability, and figure, all of which are passive qualities, nothing remains that can be interpreted as gravity or any other kind of force. "Those who assert that active force, action, and the principle of motion are really in the bodies, maintain a doctrine that is based upon no experience, and support it by obscure and

[11] "Principia vero metaphysica causaeque reales efficientes motus et existentia corporum attributorumve nullo modo ad mechanicam aut experimenta pertinent." *De motu,* sec. 41; *Works,* vol. 1, p. 515.

[12] *Siris,* sec. 232; *Works,* vol. 3, p. 232.

[13] *Siris,* sec. 231; *ibid.*

[14] "Actionem autem corporis a motu praescindere non possumus." *De motu,* sec. 11; *Works,* vol. 1, p. 505.

general terms, and do not themselves understand what they wish to say."[15]

Berkeley already recognizes a universal tendency of the human mind to transcend immediate experience and illegitimately to attribute causation to ideas. "When we perceive certain ideas of sense constantly followed by other ideas, and we know that this is not of our own doing, we forthwith attribute power and agency to the ideas themselves."[16] Similarly he says in *De motu:* "The physicist observes a succession of sense data, connected by rules, and interprets that which precedes in their order as the cause and that which follows as the effect. It is in this sense that we say that one body is the cause of the motion of another, or impresses motion on it, pulls it or pushes it."[17] Berkeley clearly realizes that the transmission of motion by impact or contact is as problematic as that by action at a distance. To introduce the term "force" as an explanatory element in the theory of physical science means to develop a misleading vocabulary. Newton's theory of mechanics and gravitation cannot be accepted as an explanation in the true sense of the word.

Mechanical laws of nature or motion direct us how to act, and teach us what to expect. Where intellect presides there will be method and order, and therefore rules, which if not stated and constant would cease to be rules. There is therefore a constancy in things, which is styled the Course of Nature. All the phenomena in nature are produced by motion. There appears an uniform working in things great and small, by attracting and repelling forces. But the particular laws of attraction and repulsion are various. Nor are we concerned at all about the forces, neither can we

[15] "Ex dictis manifestum est eos qui vim activam, actionem, motus principium, in corporibus revera inesse affirmant, sententiam nulla experientia fundatam amplecti, eamque terminis obscuris et generalibus adstruere, nec quid sibi velint satis intelligere." *Ibid.*, sec. 31, p. 511.

[16] George Berkeley, *A treatise concerning the principles of human knowledge,* sec. 32, *Works,* vol. 1, p. 274.

[17] "Physicus series sive successiones rerum sensibilium contemplatur, quibus legibus connectuntur, et quo ordine, quid praecedit tanquam causa, quid sequitur tanquam effectus, animadvertens. Atque hac ratione dicimus corpus motum esse causam motus in altero, vel ei motum imprimere, trahere etiam, aut impellere." *De motu,* sec. 7; *Works,* vol. 1, p. 527.

know or measure them otherwise than by their effects, that is to say, the motions; which motions only, and not the forces, are indeed in the bodies. Bodies are moved to or from each other, and this is performed according to different laws. The natural or mechanical philosopher endeavours to discover those laws by experiment and reasoning. But what is said of forces residing in bodies, whether attracting or repelling, is to be regarded only as a mathematical hypothesis, and not as any thing really existing in nature.[18]

Berkeley emphasizes the importance of distinguishing between a mathematical hypothesis and the nature or essence of things. Only if we are fully aware of this difference, "then all the famous theorems of mechanical philosophy which . . . make it possible to subject the world to human calculations, may be preserved; and at the same time, the study of motion will be freed from a thousand pointless trivialities and subtleties, and from (meaningless) abstract ideas." [19]

The process of reducing the phenomena to general rules, as mentioned above, may be spoken of as an explanation, if mechanical explanation as distinct from metaphysical or causal explanation is meant thereby. For we too easily confound the category of action or causation with the phenomena of motion. Too often do we conceive of force in a psychological or introspective manner, as if some creative source of action were really situated in bodies, just as the power of moving our limbs seems to be within ourselves. If we are fully cognizant of this possible misunderstanding, then the use of the term "force" may be sanctioned methodologically and may be referred to as an explanation within the limits of mechanical philosophy. Such explanation may be valid, but not real or "true." True causes in the strict sense are agents, and all agents are, according to Berkeley, incorporeal. The real cause of motion, consequently, can be only a spirit. In fact, it is the supreme efficient and final cause, it is God.

Berkeley's criticism of the concept of force may be expressed in the terminology of modern epistemology as follows: "force"

[18] *Siris,* sec. 234; *Works,* vol. 3, pp. 233–234.
[19] *De motu,* sec. 66; *Works,* vol. 1, 524.

is a construct of the conceptual scheme of physics and should not be confounded with metaphysical causality. Hume's generalization of Berkeley's ideas and, in particular, his criticism of the concept of causality itself is a straightforward continuation of Berkeley's inception. It is not only of historical importance to note that the critical attitude toward the concept of force and causation in mechanics had been attained several decades before the publication of Hume's well-known exposition of this question. However, Hume's work undoubtedly contributed much to a further elucidation of the matter. Pierre Louis Moreau de Maupertuis, for example, was strongly influenced by Hume in his concept of force and causation. For the philosophically minded of that generation it became more and more obvious that the scientific notion of force has little to do with causal explanation, and that, moreover, the connection between "cause" and "effect" in science, or antecedent and sequent in the succession of phenomena, is not a matter of logical inquiry, but of experimental experience or observation. Logical analysis is powerless and inadequate to explain transfer of motion in either case, whether force is conceived as an action of contact or as an action at a distance. The two concepts face equal logical and metaphysical difficulties, since both of them are nothing but constructs, descriptive names of perceptible and measurable empirical relations. It is pure prejudice to assume that action at contiguity is more intelligible and more rational than action at a distance. Maupertuis, in his *Essai de cosmologie,* discusses this point in detail.

> If someone who had never touched a body or seen how bodies collide, but who was experienced in mixing colors, saw a blue object move toward a yellow one, and were asked what would happen if these two bodies collide, he would most probably say that the blue body would turn green as soon as it united with the yellow one.[20]

[20] "Si quelqu 'un qui n'eût jamais touché de corps, et qui n'en eût jamais vu se choquer, mais qui eût l'expérience de ce qui arrive lorsqu'on mêle ensemble differentes couleurs, voyoit un corps bleu se mouvoir vers un corps jaune, et qu'il fût interrogé sur ce qui arrivera lorsque les deux corps se rencontreront;

Under no circumstances, however, can he predict that both will move after the encounter with a common velocity, or that the one will communicate to the other part of its velocity or rebound from it. It is only our tactile experience with common objects that leads us to the concept of impenetrability and enables us to formulate the rules or laws of impact, without affording the least information about the real occurrence behind the phenomenon.

We speak of forces, declares Maupertuis, only to conceal our ignorance "un mot qui ne sert qu'à cacher notre ignorance." And he continues with a statement that might well serve as the motto of the present book:

Il n'y a dans la philosophie moderne aucun mot répété plus souvent que celui-ci, aucun qui soit si peu exactement défini. Son obscurité l'a rendu si commode, qu'on n'en a pas borné l'usage aux corps que nous connoissons; une école entière de philosophes attribue aujourd'hui à des êtres qu'elle n'a jamais vus une force qui ne se manifeste par aucun phénomène.[21]

It is in analogy to the sensation we have when moving an object from its place or arresting the motion of another that we ascribe a similar state of affairs to the phenomena of physical motion. However, says Maupertuis, since we cannot emancipate ourselves completely from the idea that bodies exert mutual influences upon each other, we may continue to use the term "force." But we should always remember that the concept of force is but an invention to satisfy our desire for explanation.[22]

It is not easy to comply consistently with this rule. Even Newton, in spite of his critical vigilance and circumspection, fell a victim to this trap, as Maupertuis declares in his *Examen*

---

peut-être que ce qu'il pourroit dire de plus vraisemblable seroit, que le corps bleu deviendra verd dès qu'il aura atteint le corps jaune." P. L. M. de Maupertuis, *Essai de cosmologie,* in *Oeuvres* (Lyons, 1756), vol. 1, p. 31.

[21] *Ibid.*, pp. 28–29.

[22] "Et nous nous souviendrons toujours que la force motrice, la puissance qu'a un corps en mouvement d'en mouvoir d'autres, n'est qu'un mot inventé pour suppléer à nos connoissances, et qui ne signifie qu'un résultat des phénomènes." *Ibid.*, p. 31.

philosophique.[23] For Newton's second law of motion, stating that the change of motion is proportional to the impressed force, is, strictly speaking, in Maupertuis' view, only a definition of force (an empty tautology), but was regarded by Newton as an important law of nature.[24]

It is historically interesting to note that Maupertuis' discussion of impact phenomena appeared in the publications of the Berlin Academy in 1746, that is, at about the time when Boscovich and Kant were investigating the same problem. David Hume's *Enquiry concerning human understanding* appeared only in 1748, in London, but his *Treatise of human nature* was already published there in 1739–1740. At least as far as Maupertuis and Kant are concerned, there is no doubt that Hume's investigation respecting the origin and application of the idea of causality had its decisive influence on these investigations. As is well known, Hume's thesis that the origin of the concept of causality is found in habit reduced causality to a mere association of perceptions. In fact, says Hume, we have "no other notion of cause and effect but that of certain objects which have been conjoined together habitually in past experience," and the possibility of understanding the nature of an objective connection between cause and effect must be denied. For such a connection can be discovered neither by rational insight (since no logical contradiction would be involved by assuming the cause without the effect or vice versa) nor by observation (since forces of production have no sensible qualities). Our attempt to transcend by means of causal ideas the domain of experience is philosophically illegitimate. What Maupertuis did was simply to apply these ideas to the most fundamental phenomenon in mechanics, the impact of two bodies, a phenomenon which in Cartesian physics and in the mechanistic theories of gravitation was conceived as an unproblematic and

[23] *Examen philosophique de la preuve de l'existence de Dieu,* Deuxième partie, secs. xxiii and xxxvi.

[24] If Maupertuis had known Kant, he would certainly have added: "and synthetic proposition."

self-evident mechanism for the transmission of force, based on contiguous action.

With the investigations of Boscovich, Kant, and Maupertuis, it became clear that the contiguous phenomenon of impact fundamentally was not less problematic than the concept of an action at a distance. It was no longer a question of the precise mode of propagation of force, but the very Newtonian concept of force itself became the object of critical analysis. Yet, for Boscovich as well as for Maupertuis, the question why a certain body moves, what is the origin of its motion, had still an answer. It was the answer given already by Berkeley: God.

Hume's criticism of causation became influential again in the second half of the nineteenth century when theological reasoning was no longer being accepted as relevant for argumentations in physics. Hume's criticism of causality, thus, formed the foundation of the more or less positivistic doctrines in mechanics and physics as advanced by Kirchhoff, Mach, Hertz, Poincaré, Russell, and others, for whom the concept of force was but the physical equivalent of Hume's conception of "necessary connection" between cause and effect, an idea illegitimate in science.

The empirical, antimetaphysical attitude in mechanics, culminating toward the end of the nineteenth century in the attempts of Kirchhoff, Hertz, and Mach to eliminate the concept of force from science, was not a sudden novelty, unprepared and unexpected in the development of scientific thought. As we have seen in the previous chapter, this change in the climate of opinion can be traced back to Berkeley's criticism of Newton's mechanics and to the Humean analysis of the concept of causality. So far, only the more general, philosophical aspects of this development have been discussed, but now we shall turn to its effect on physics proper.

Leonhard Euler, in his *Mechanica*,[25] proposed to construct, by means of axioms, definitions, and logical deductions, a rational

[25] Leonhard Euler, *Mechanica, sive motus scientia analytice exposita* (St. Petersburg, 1736).

science of mechanics and to demonstrate that Newtonian mechanics is an apodictic science, not of contingent but of necessary truth. Still adopting the standpoint of classical realism, Euler takes force as the fundamental concept in his dynamics, although the comparison and measurement of forces is relegated to statics. Definition I of his deductive system states: "Potentia est vis corpus vel ex quiete in motum perducens, vel motum ejus alterans." ("Potency is a force which initiates the motion of a body [originally] at rest or which alters its motions.") Gravity, for example, is such a force and therefore a potency. Euler's "potency" (*potentia*) corresponds to what was generally called "accelerative force" (*vis accelerativa*) and in the following will be denoted simply by "force." In Scholium II Euler discusses how statics is concerned with the measurement of these forces. Force A corresponds to force B as (the number) m to n, if A, applied n times in a certain direction on a given point, and B, applied m times on the same point in the opposite direction, leaves the point in equilibrium. It is an axiom of Euler's conceptual scheme that forces whose equivalence has been proved by statics exhibit also the same dynamic effects. Operationally viewed, statics supplies the foundation of dynamics, although the latter, according to Euler, forms the basis for a deductive theory of mechanics. In contrast to Varignon, who in his *Nouvelle mécanique* bases his theory on statics, Euler, in his *Mechanica sive motus scientia,* reduces theoretical mechanics to dynamics. Euler then distinguishes between absolute forces, such as gravity, whose dynamic effects are independent of whether the body affected is at rest or in motion, and relative forces, whose effects depend on the velocity of the body, as the hydrodynamic force of a liquid current on an object. At first Euler discussed the effect of a force upon only one body (*corpus* means material point or particle). Then, in order to generalize the results for a set of bodies, he introduces the concept of mass and thus completes the basis for his subsequent development of dynamics.

Jean Le Rond d'Alembert, in his *Traité de dynamique,* embraces a more empirical point of view. He claims that all that we perceive distinctly in the motion of a body is the fact that it traverses a certain distance in a certain time.

C'est donc de cette seule idée qu'on doit tirer tous les Principes de la Mécanique, quand on veut les démontrer d'une manière nette et précise; ainsi on ne sera point surpris qu'en conséquence de cette réflexion j'aie, pour ainsi dire, détourné la vue de dessus les causes motrices pour n'envisager uniquement que le mouvement qu'elles produisent.[26]

D'Alembert consciously ignores causes of motion as being completely unknown to us. The only exception is impact or impulse, as he calls it.[27] All other causes are only indirectly ascertainable through their effects of accelerating or retarding the motion of bodies.[28] Consequently, d'Alembert introduces what he calls "accelerative force" (*force accélératrice*), ϕ, by the equation

$$\phi\, dt = du,$$

in which dt and du signify infinitesimal increments of time and velocity. Force, thus, attains the status of a derived notion in d'Alembert's system. In consonance with Newton's terminology (see page 122), d'Alembert defines "motive force" (*force motrice*) as the product of accelerative force and mass. D'Alembert seems to be unable to supply a strict definition of mass, for he adopts this concept as a priori known, which is of course a serious deficiency in the logic of his system. The same difficulty was encountered by most systematizations of mechanics prior to Mach, which defined the concept of force (or motive force) as the product of acceleration (corresponding roughly to d'Alembert's accelerative force) and mass, while those systematizations

[26] J. L. d'Alembert, *Traité de dynamique* (Paris, 1743), p. xvi.

[27] "Or, de toutes les causes, soit occasionelles, soit immédiates, qui influent dans le mouvement des corps, il n'y a tout au plus que l'impulsion seule dont nous soyons en état de déterminer l'effet par la seule connaissance de la cause"; *ibid.*, p. 16.

[28] "Elles ne peuvent par conséquent se manifester à nous que par l'effet qu'elles produisent en accélérant ou retardant le mouvement des corps, et nous ne pouvons les distinguer les unes des autres que par la loi et la grandeur connue de leurs effets, c'est-à-dire par la loi et la quantité de la variation qu'elles produisent dans le mouvement." *Ibid.*, pp. 16–17.

that introduced force as a primitive notion derived mass as the quotient of force and acceleration. Either force or mass, in short, was taken as an intuitive notion.

In 1788 Louis de Lagrange's monumental *Mécanique analytique* was published, with the purpose of bringing Newtonian mechanics to its mathematical perfection. Lagrange is less interested in the methodological aspects of the fundamental concepts than in their contribution to a mathematical unification of the science. This is clearly stated in the "Avertissement" to the work: "Réduire la théorie da la mécanique et l'art d'y résoudre les problèmes qui s'y rapportent à des formules générales, dont le simple développement donne toutes les équations nécessaires pour la solution de chaque problème." Mass times accelerative force (acceleration) is his formula for "force élémentaire ou naissante"; this quantity, if considered as the measure of the effort that a body can exert by virtue of the velocity possessed or intended, constitutes what is called pressure; but if considered as the measure of the force necessary to impress the same velocity, it is called motive force (*force motrice*). Having thus introduced the dynamic concepts of his system, he continues the construction of rational mechanics along purely mathematical lines.

The first systematic exposition of mechanics to be strongly influenced by British empirical philosophy is that of Lazare Carnot, father of Sadi Carnot. Lazare Carnot's first treatise on mechanics, *Essai sur les machines en général,* appeared in Paris in 1783, and was published, enlarged and revised, under the title *Principes fondamentaux de l'équilibre et du mouvement* in Paris in 1803. We have already referred to these works in connection with Newton's concept of force as a sequence of impulses. But before discussing this point in detail, let us examine the general climate of opinion on which Lazare Carnot's treatises are based. In the preface to his *Principes* Carnot admits that there are two different approaches to the study of mechanics; in one the subject is regarded as a theory of forces as the causes of motion, and in the other mechanics is interpreted as the theory of motion itself. The first approach, according to Carnot, is the one generally adopted for instruction, as being the simpler. But it has the disadvantage of being based on an obscure metaphysical notion, that of force. "Mais elle a le désavantage d'être fondée sur une notion métaphysique et obscure qui est celle des forces. Car quelle idée nette peut présenter à l'esprit en pareille matière le nom de cause? il y a tant d'espèces de cause! Et que peut-on entendre dans le langage précis des Mathématiques par une force, c'est-à-dire par une cause double ou triple d'une autre?" In order to avoid the metaphysical notion of force and not to distinguish between cause and effect, Carnot prefers the second method of approach. Unfortunately, however, his treatment is not always strictly consistent and his style often diffuse and exuberant. Furthermore, his criticism of the metaphysical nature of the concept of force is not applied with respect to the concept of mass, which, like d'Alembert, he introduces as an intuitive notion. In fact, even while emphasizing his antimetaphysical attitude with reference to the concept of force, he speaks of mass without recognizing that this concept presents a similar difficulty. He says, "Ainsi les forces, telles qu'on les considère en Mécanique, ne sont pas des êtres métaphysiques et abstraits: chacune d'elles réside dans une masse déterminé." [29] Having now at his disposal the concept of mass, he defines quantity of motion in the conventional manner as the product of mass and velocity: mv. Finally, in consonance with his conception of force as a series of infinitesimal increments of impulses, he introduces the concept in a way that corresponds to the well-known equation $F\,dt = d(mv)$.

As the last representative of this French school of mechanics we mention Barré de Saint-Venant, to whom we have already repeatedly referred. His ideas show a striking similarity to those of Carnot. In a paper published by the scientific society of Lille, Saint-Venant makes the important statement that whatever problem in terrestrial or celestial mechanics we consider,

[29] L. N. M. Carnot, *Principes fondamentaux de l'équilibre et du mouvement* (Paris, 1803), p. 47.

forces never appear among the data or in the answer to the problem. They are introduced merely as auxiliary concepts to facilitate the solution of the problem and eventually have to be eliminated in order to state the answer in terms of distances, times, and velocities.[30] So far Saint-Venant only reaffirms the ideas of d'Alembert and Carnot, though with much more precision and lucidity. But now, his original contribution is the clear recognition that the introduction of the concept of force as a derived notion presupposes a strict definition of the concept of mass. Consequently, as expounded in a paper published in the *Comptes rendus* in 1845[31] and in a short elementary treatise, *Principes de mécanique fondés sur la cinématique* (Paris, 1851), a strict definition of the concept of mass precedes that of force. The general idea of his definition of mass, although based on his fundamental assumption of atomism, makes him a precursor of Ernst Mach. Saint-Venant defines mass thus: The mass of a body is the ratio of two numbers which express how often the body and a standard body contain parts which, if separated and then brought into mutual collision, two by two, communicate to each other opposite equal velocities.[32] "From the practical point of view" ("à notre point de vue tout pratique"), it matters very little, says Saint-Venant, whether the concept of mass, thus defined, has any relation to what is generally called "quantity of matter," just as it is unimportant whether the concept of force (which he defines essentially as the product of mass and accelera-

[30] "Dans le fait, quel que soit un problème de Mécanique terrestre ou céleste proposé, les forces n'entrent jamais ni dans ses données, qui sont toujours des choses sensibles, ni dans le résultat cherché de la solution. On les fait intervenir pour résoudre, et on les élimine ensuite afin de n'avoir finalement que des temps et des distances ou des vitesses comme en commençant." A. J. C. Barré de Saint-Venant, "Notice sur Louis-Joseph comte du Buat," *Mémoires de la Société des sciences de Lille* (1865).

[31] "La masse d'un corps est le rapport de deux nombres exprimant combien de fois ce corps et un autre corps, choisi arbitrairement et constamment le même, contiennent de parties qui, étant séparées et heurtées deux à deux l'une contre l'autre, se communiquent, par le choc, des vitesses opposées égales." Barré de Saint-Venant, "Mémoire sur les sommes et les différences géométriques et sur leur usage pour simplifier la mécanique," *Comptes rendus 21,* 620 (1845).

[32] *Ibid.,* p. 624.

tion) has any relation to the notion of efficient causes of motion. One day, he hopes, physics will not have any need of such intermediate notions of an occult and metaphysical nature, but will proceed to the solution of its problems by the application of laws based solely on concepts of velocities and their changes according to the circumstances.[33] Saint-Venant thus inaugurated the program that was to be carried out, about thirty years later, by Mach, Kirchhoff, and Hertz.

Before discussing this last stage in the gradual process of eliminating the concept of force from classical physics, we have to digress for a moment to take notice of a remarkable revival of Euler's doctrine of the primacy of force. In 1852, Ferdinand Reech, in his *Cours de mécanique*,[34] laid the foundation of what may be called "the school of the thread." The French call it "école du fil." In opposition to d'Alembert and Saint-Venant, Reech and his most important disciple, Andrade,[35] took as the starting point for their exposition of mechanics the concept of force — not force in the general comprehensive meaning, but force almost in the Aristotelian sense of push or pull. For, they contend, the only true and genuine notion of force that we can form on the basis of immediate experience is that of pressure or traction. "Nous éprouvons alors des sensations qui éveillent en nous plusieurs idées fondamentales: d'abord celle de l'existence des corps, puis celle de la forme des corps et des propriétés de l'espace, puis celle du mouvement et du temps, puis encore celle d'une certaine quantité que nous nommons une pression ou une traction."[36] When acting on objects of different sizes or of equal sizes but of different materials, experience shows that equal

[33] "On concoit très bien qu'un jour, à la place de ces sortes d'intermédiaires d'une nature occulte et métaphysique, on puisse n'introduire et n'invoquer, pour la solution des divers problèmes de l'ordre physique, que les lois avérées des vitesses et de leurs changements suivant les circonstances." Barré de Saint-Venant, "Notice sur Louis-Joseph."

[34] Ferdinand Reech, *Cours de mécanique d'après la nature généralement flexible et élastique des corps* (Paris, 1852).

[35] J. F. C. Andrade, *Leçons de mécanique physique* (Paris, 1897).

[36] Reech, *Cours de mécanique*, sec. 1.

kinematic changes require unequal forces. Thus we are led to the assumption that the force is proportional to the number of identical particles that constitute the object. In the case of bodies made of different materials, two parts, one of each, are equivalent as to their material qualities, if the same force produces in them kinematic changes. Thus we arrive at the concept of mass. Now, every physical body is characterized not only by mass, but also by the cohesion ("liaison") of its parts. As experience shows, cohesion can be employed experimentally for the determination of force, or rather of the intensity of force. "L'expérience nous apprend aussi que la rupture ou la séparation du système de liaison d'un corps n'a lieu qu'après un certain changement de figure, préalablement accompli, et par là nous comprenons que la qualité liaison des corps pourra nous servir à trouver expérimentalement les intensités des forces dès le début et avant l'établissement d'aucune science mécanique." [37]

On these presuppositions Reech now develops, in a peculiar combination of theory and experience, his mechanics. The fundamental dynamical phenomenon is the elongation of an elastic thread or string, which is not only the indication for the existence of a force, acting in the string on the object attached at its end, but also the dynamometric measure of the intensity of the force. It is then easy to show how forces can be compared and measured relative to each other. So far Reech's method is not much different from an ordinary introduction to statics. The difficulty only arises when dynamical phenomena have to be accounted for and when gravitation and electrostatic and magnetic attractions or repulsions have to be considered. Strictly speaking, concepts like attractions or repulsions have no place in the "dynamics of threads." In fact, Reech refers to such phenomena as "causes mystérieuses agissantes." In our outline of his program we shall call these latter forces, for the sake of brevity, "field forces."

In order to cope with the most general conditions, Reech

[37] *Ibid.*, sec. 8.

considers a system composed of an elongated string with a body (particle) attached to one of its ends, moving in space, and he takes it as then being in a state of equilibrium between the force **F**, acting in the string, on the one hand, and the inertial force exerted by the body on the other. As in Huygens's treatment of centrifugal forces, Reech assumes that the string is suddenly cut. The acceleration of the body will then undergo a discontinuity $\mathbf{f} = \mathbf{a} - \mathbf{b}$. From experience **b** can be known. Its value depends on the "field forces." Reech now shows how this change of acceleration **f** has to be identified with the acceleration of the inertial force, arriving thus at the final law of motion $\mathbf{F} = m\,\mathbf{f}$. For this purpose he assumes, guided by experience, that **F** is proportional to the mass m, and has the direction of **f**. He thus arrives at a vectorial equation,

$$\mathbf{F} = m\mathbf{f}\,\psi(f, \ldots),$$

in which ψ is a scalar function possibly of f, of the position, and of the time. On grounds of the isotropy of space, the homogeneity of time, and the independence of the effects of simultaneous forces, Reech tries to demonstrate that the function ψ is independent of position and time, and finally must be a constant, which through a proper choice of units can be taken as 1. Summarizing the fundamental formula $\mathbf{F} = m(\mathbf{a} - \mathbf{b})$ Reech comments:

> On voit que ce ne sera pas une de ces vérités élémentaires qu'on puisse admettre en guise d'axiome dès le début de la science. Ce ne sera pas non plus une vérité purement abstraite, ni purement expérimentale; il s'y trouvera beaucoup de l'une et beaucoup de l'autre, et les applications ultérieures de la science qui en dépendra devront en vérifier la parfaite justesse.[38]

As Reech himself admits in these words, the logical structure of his system of mechanics is rather complicated. Furthermore, although a reinstatement of the notion of force, his system nevertheless shows strong antimetaphysical tendencies. In fact,

[38] *Ibid.*, sec. 34.

the term force does not denote any cause of motion, according to this view, but is almost synonymous with the elastic deformation originated by what is commonly called pressure or traction. "Le mot force ne devant plus servir à désigner une cause quelconque de mouvement, mais . . . cet effet particulier d'une cause quelconque qu'on nomme une pression ou une traction et que nous apprécions avec un si haut degré de clarté dans un fil tendu." [39] Ultimately, it seems, the exponents of the "dynamics of strings" tried to construct their system of mechanics from an operational point of view, but were not yet in a position to carry out their program consistently.

More successful in this respect were those theories of mechanics that divested themselves of the concepts of cause and force (and finally also of the concept of substance) and adopted the purely functional point of view, taking force as a derived concept devoid of all temporal, causal, or teleological implications. In fact, force was only a name to signify the product of mass and acceleration, and any other name, from this point of view, would have done as well. The chief exponent of this doctrine, which may be called deanthropomorphic, positivistic, or operational, was Ernst Mach. His first statement along these lines was a brief essay of five octavo pages "On the definition of mass," [40] which was rejected by Poggendorf's *Annalen* in 1867 and was published one year later in Carl's *Repertorium der Experimentalphysik*. Referring to three experimental facts, enunciated in what Mach calls "experimental propositions," the foundations of Newtonian mechanics are reconstructed with the help of two definitions as follows:

First Experimental Proposition: Bodies induce in each other contrary accelerations in the direction of their line of junction.

[39] *Ibid.*, sec. 2, Les principes de la dynamique. For a similar conception, cf. Joseph H. Keenan, "Definitions and principles of dynamics," *Scientific Monthly* 67, 406 (1948).

[40] Ernst Mach, "Ueber die Definition der Masse," Carl's *Repertorium der Experimentalphysik* 4 (1868); reprinted in Mach's *Erhaltung der Arbeit* (Leipzig, second ed., 1909).

First Definition: The negative inverse ratio of the mutually induced accelerations of any two bodies is called their mass ratio.

Second Experimental Proposition: These mass ratios are independent of the character of the physical states of the bodies.

Third Experimental Proposition: The accelerations induced by a number of bodies in a given body are independent of each other.

Second Definition: The product of the mass value and the acceleration induced in that body is called the moving force. (Here the mass value is of course the mass ratio of the body with reference to an arbitrarily chosen standard body of unit mass.)

Force, as well as mass, are thus reduced in Mach's conception of mechanics to purely mathematical expressions relating certain measurements of space and time. A close analysis of the statements outlined above may show that further experimental evidence has to be adduced for a completely logical construction of classical mechanics. Thus, for example, it seems necessary to assume the following additional experimental proposition: the ratio of the accelerations induced mutually by two arbitrary bodies A and B is equal to the ratio of the accelerations induced by A and an arbitrary third body C divided by the ratio of the accelerations induced by B and C. However, such modifications are of minor importance and do not depreciate Mach's contribution to the attempt to regard the abstract determinative element in a physical situation as a purely mathematical function of the arbitrarily given constellation of bodies. This replacement of the concepts of cause and force by mathematical functions or relational concepts was, in his view, not only the program of mechanics but that of science as a whole. "I hope," he declared in his *Popular scientific lectures,* "that the science of the future will discard the idea of cause and effect, as being formally obscure; and in my feeling that these ideas contain a strong tincture of fetishism, I am certainly not alone." [41]

[41] Ernst Mach, *Popular scientific lectures,* trans. T. J. McCormack (Open Court, La Salle, Ill., 1943), p. 254.

Indeed, Mach was certainly not alone in his conceptions. In Germany Gustav Kirchhoff and Heinrich Hertz, in England William Kingdon Clifford and Karl Pearson, advanced similar views. Kirchhoff's *Lectures on mechanics* appeared in 1876. Its introductory lines have often been quoted as marking an era in scientific thought: "Mechanics is the science of motion; its task is defined as follows: to describe completely and in the simplest manner the motions which take place in nature."[42] In confining itself to description, mechanics, in the view of Kirchhoff, has to discard the search after causes. The necessary elements for such a description are the concepts of space, time, and matter.[43] In contrast to Mach's exposition, Kirchhoff's mechanics adopts the concept of mass, as intuitively given, on the same footing as the concepts of space and time. In fact, masses are introduced as constant coefficients characteristic of the mass particles under consideration. Experience shows that the introduction of these coefficients is useful for a complete description of the motions. Force is then defined as the product of mass and acceleration and is thus a generalization of what Kirchhoff calls "accelerative forces" X, Y, and Z, defined by the equations:

$$\frac{d^2x}{dt^2} = X, \quad \frac{d^2y}{dt^2} = Y, \quad \frac{d^2z}{dt^2} = Z.$$

Since "force" has no meaning other than as a name for the second derivative of the coördinates with respect to time, if we ignore the constant coefficient denoting the mass of the body, a new legitimate problem arises: why just the second derivatives and why not the third or fourth? That the first derivative had to be excluded from these considerations was obvious by the law of inertia. As to the higher derivatives, Kirchhoff, in fact, raises this question in section 4 of his *Lectures*. Experience teaches us,

[42] "Die Mechanik ist die Wissenschaft von der Bewegung; als ihre Aufgabe bezeichnen wir: die in der Natur vor sich gehenden Bewegungen vollständig und auf die einfachste Weise zu beschreiben." Gustav Kirchhoff, *Vorlesungen über Mechanik* (Leipzig, 1876; fourth ed., 1897), p. 1.

[43] "Zur Auffassung einer Bewegung sind die Vorstellungen von Raum, Zeit und Materie nöthig, aber auch hinreichend." *Ibid.*

he contends, that the motions which take place in nature are of such a kind that their representation by higher derivatives than the second would contribute nothing to the simplification of their description; on the contrary, it would merely complicate the description, the reason being, as experience shows, that the second derivatives of the coördinates of mass particles with respect to time are themselves functions of the coördinates.[44] For the sake of historical completeness it should, however, be noticed that attempts to introduce forces of higher order, corresponding to the third and fourth derivatives of the coördinates, have been made by C. Neumann, W. Voigt, and L. Koenigsberger.

A material point is submitted to a system of forces, according to Kirchhoff, if its motion conforms to the equations

$$\frac{d^2x}{dt^2} = X_1 + X_2 + \cdots ; \frac{d^2y}{dt^2} = Y_1 + Y_2 + \cdots ; \frac{d^2z}{dt^2} = Z_1 + Z_2 + \cdots .$$

In case the system comprises only two forces, these equations are the analytical expression for the so-called theorem of the parallelogram of forces. If the motion of a material point, continues Kirchhoff, is conditioned by various forces, these forces cannot separately be determined; only their resultant is determined. All forces (X_1, Y_1, Z_1), (X_2, Y_2, Z_2), . . . can be chosen arbitrarily with the exception of one single force, which then always can be taken in such a manner that the final resultant is equal to the acceleration given in experience. Now, since motion alone is the source of concept formation in mechanics, claims Kirchhoff, mechanics cannot supply a complete definition of the concept of force.

Es folgt daraus, dass nach Einführung von Kräftesystemen an Stelle einfacher Kräfte die Mechanik ausser Stande ist, eine vollständige Definition des Begriffs der Kraft zu geben. Trotzdem ist diese Einführung von der höchsten Wichtigkeit. Es beruht das darauf, dass, wie die Erfahrung gezeigt hat, bei den natürlichen Bewegungen sich immer solche Systeme finden lassen, deren Einzelkräfte leichter angegeben werden können, als ihre Resultanten.[45]

[44] *Ibid.*, p. 6.
[45] *Ibid.*, p. 11.

Kirchhoff's exposition of mechanics was followed by Heinrich Rudolf Hertz's *Principles of mechanics*,[46] published posthumously in 1894. Hertz's conceptions of mechanics was strongly influenced by his work in electrodynamics. His aversion to the traditional concept of force originated in his study of the theory of electrodynamics as advanced by Faraday, Maxwell, and Helmholtz. In particular, it was the mathematical theory of the transmission of electrodynamic action as formulated by Gauss, Riemann, Betti, and C. Neumann that called for a mechanical representation as the aim of many theoreticians of electrodynamics. As James Clerk Maxwell says in his *Treatise on electricity and magnetism:* "Hence all these theories lead to the conception of a medium in which the propagation takes place, and if we admit this medium as a hypothesis, I think it ought to occupy a prominent place in our investigations, and that we ought to endeavour to construct a mental representation of all the details of its action, and this has been my constant aim in this treatise."[47] Just as Maxwell conceived electromagnetic forces as due to the motion of concealed masses, or as Lord Kelvin reduced these effects to a mechanism of vortex atoms and Helmholtz to cyclical systems of concealed motion, so Hertz thought it necessary to account not only for electrodynamic forces, but also for gravitational forces, for all actions at a distance, and finally for all mechanical forces, by some mechanism of concealed masses and motions. But if such an approach is capable of gradually eliminating the mysterious forces from mechanics, declares Hertz, it should be possible entirely to prevent their entering into mechanics.

In the introduction to his *Principles of mechanics* Hertz criticizes the traditional conception of force in Newtonian

[46] Heinrich Hertz, *Die Prinzipien der Mechanik in neuem Zusammenhang dargestellt* (Leipzig, 1894); *The principles of mechanics presented in a new form,* trans. D. E. Jones and J. T. Walley (London, 1899); with a new introduction by R. S. Cohen (Dover, New York, 1956).

[47] James Clerk Maxwell, *A treatise on electricity and magnetism* (Oxford, 1873), vol. 2, p. 438; third ed. (1892), p. 493.

mechanics. For this purpose he cites the example of a stone tied to a string and whirled around in a circular path. We are conscious, he states, of exerting a force upon the stone. This force now, deflects the stone constantly from its straight path. By inserting a dynamometer we can easily convince ourselves of the correctness of Newton's second law of motion for this case.

> But now the third law requires an opposing force to the force exerted by the hand upon the stone. With regard to this opposing force the usual explanation is that the stone reacts upon the hand in consequence of centrifugal force, and that this centrifugal force is in fact exactly equal and opposite to that which we exert. Now is this mode of expression permissible? Is what we call centrifugal force anything else than the inertia of the stone? Can we, without destroying the clearness of our conceptions, take the effect of inertia twice into account, — firstly as mass, secondly as force? In our laws of motion, force was a cause of motion, and was present *before* the motion. Can we, without confusing our ideas, suddenly begin to speak of forces which arise through motion, which are a consequence of motion? Can we behave as if we had already asserted anything about forces of this new kind in our laws, as if by calling them forces we could invest them with the properties of forces? These questions must clearly be answered in the negative.[48]

Hertz thus comes to the conclusion that what is commonly called centrifugal force is, properly speaking, not a force at all. It is in his view merely a matter of historic tradition that we denote it with the name of a force. But now he asks: "What becomes of the demands of the third law, which requires a force exerted by the inert stone upon the hand, and which can only be satisfied by an actual force, not a mere name?" It is unfortunate that Hertz did not clarify his ideas through the use of the concept of inertial forces. But his criticism is deeper than this. He seems to be justified in interpreting the force spoken of in the definition and in the first two laws as acting upon a body in one definite direction, whereas in his view the third law conceives forces as always connecting two bodies and as directed from the first to the second body as well as from the second to the first. He

[48] Hertz, *Principles of mechanics,* p. 6.

therefore declares: "It seems to me that the conceptions of force assumed and created in us by the third law on the one hand, and the first two laws on the other hand, are slightly different."

Furthermore, Hertz criticizes traditional mechanics from the standpoint of logical simplicity. Is it sparing in unessential characteristics? he asks. In answering this question he turns again to the idea of force. "It cannot be denied that in very many cases the forces which are used in mechanics for treating physical problems are simply sleeping partners, which keep out of the business altogether when actual facts have to be represented." [49] Thus, although in simple mechanical relations, as when lifting a stone, the force exerted (by the arm) seems to be readily and directly perceptible, in more advanced problems, as in celestial mechanics, force never is an object of direct perception. It is a transitory aid in the calculation and disappears finally from our considerations. The same applies to chemical forces, molecular forces, and many electric and magnetic actions.

Hertz subsequently criticizes any representation of mechanics based on Hamilton's principle or any other equivalent principle (variational principle). Apart from its limited applicability to holonomous systems — Hertz quotes the case of a sphere rolling without slipping upon a horizontal plane under the influence of inertia alone as not covered by Hamilton's principle — such a representation has serious flaws in its logical permissibility. Its fundamental concepts are space, time, mass, and energy; but for the last concept no clear definition can be given without falling back on the traditional formulation of mechanics.

To avoid all these insufficiencies Hertz now proposes his new arrangement of the principles of mechanics, starting with only three independent fundamental conceptions: space, time, and mass. The problem he has to solve is to represent the relations among these three, and among these three alone. Force, as an independent autonomous conception, is avoided. To this point

[49] *Ibid.*, p. 11.

Hertz's mechanics shows great resemblance to Kirchhoff's conceptions; Kirchhoff also began his exposition with the exclusive use of the concepts of space, time, and mass. But now, in contrast to Kirchhoff (for whom forces were represented only by their kinematic effects as accelerations), Hertz, in order to obtain "an image of the universe which shall be well-rounded, complete, and conformable to law," presupposes other, invisible things behind the things that we see, "confederates concealed beyond the limits of our sense." For Newtonian mechanics these were the forces; for Hamiltonian mechanics it was energy. Hertz, for the sake of logical simplicity, assumes that this hidden something is nothing else than motion and mass again, thereby denying a special category to which this something belongs. He thus proposes the hypothesis that it is possible "to conjoin with the visible masses of the universe other masses obeying the same laws, and of such a kind that the whole thereby becomes intelligible and conformable to law." Between the various concealed masses certain kinematic relations or connections are postulated, conditioning their motions and expressible by homogeneous linear differential equations of the first order, thereby satisfying the conditions of continuity. The behavior of the system as a whole, as well as of its individual constituents, can be epitomized by one single fundamental law. By a skillful and suitable definition of the expressions involved, this law can be enunciated as follows: Every natural motion of an independent material system is such that the system follows with uniform velocity the path of minimum curvature. This law becomes intelligible, of course, only if its concepts, such as "independent," "velocity of a system," "straightest path," are defined mathematically. The aggregate of the positions that a system occupies in its passage from one position to another is called a path of the system. The path is analytically represented when the coördinates of its positions are given as functions of any one chosen variable. It is straight if it has the same direction in all its positions. If M denotes the mass of the system, that is, $M = \Sigma m_i$, the length S of a path is defined by the equation[50]

$$MS^2 = \Sigma[m(dx^2 + dy^2 + dz^2)],$$

and its velocity by

$$V = \frac{dS}{dt}.$$

The rate of change of the direction of a path with respect to the length of the path is called the curvature of the path.

After these preliminary definitions, we can easily deduce the analytic expression of Hertz's fundamental law, if we assume, for the sake of brevity, Gauss's principle of least constraint:[51]

$$\Sigma\left\{m\left[\frac{d^2x}{dt^2}\,\delta\left(\frac{d^2x}{dt^2}\right) + \frac{d^2y}{dt^2}\,\delta\left(\frac{d^2y}{dt^2}\right) + \frac{d^2z}{dt^2}\,\delta\left(\frac{d^2z}{dt^2}\right)\right]\right\} = 0,$$

in which the accelerations are varied without variation of the coördinates of position x, y, z, . . . or of the velocities dx/dt, . . . If, now, x, y, and z are considered as functions of S, and if derivatives with respect to S are marked by primes, it is obvious that

$$\frac{dx}{dt} = x'V, \ldots \qquad \frac{d^2x}{dt^2} = x''V^2, \ldots$$

and Gauss's principle can be written

$$\delta\Sigma\{m(x''^2 + y''^2 + z''^2)\} = 0.$$

Hertz now defines the curvature c of a path by the equation[52]

$$Mc^2 = \Sigma[m(x''^2 + y''^2 + z''^2)],$$

and his law of motion, requiring minimum curvature, reduces evidently to Gauss's principle. Moreover, it is clear that V is constant since

$$V = \frac{dS}{dt} = \left[\frac{2(\text{kinetic energy})}{M}\right]^{\frac{1}{2}}.$$

[50] *Ibid.*, p. 73.

[51] C. F. Gauss, "Ueber ein neues Grundgesetz der Mechanik," *Journal de Crelle* 4 (1829).

[52] Hertz, *Principles of mechanics*, p. 75.

From this law, together with the hypothesis of concealed masses and normal connections, the rest of Hertz's mechanics can be derived by purely deductive reasoning. Although mechanics is thus in point of principle reduced to kinematics, Hertz nevertheless finds it convenient to introduce into his system the concept of force. "However, it is not as something independent of us and apart from us that force now makes its appearance, but as a mathematical aid whose properties are entirely in our power. It cannot, therefore, in itself have anything mysterious to us." According to Hertz's fundamental law of motion, the motion of either one of two bodies belonging to the same system is determined by the motion of the other. Hertz finds it convenient to divide the determination of the one motion by the other into two steps. "We thus say that the motion of the first body determines a force, and that this force then determines the motion of the second body. In this way force can with equal justice be regarded as being always a cause of motion, and at the same time a consequence of motion. Strictly speaking, it is a middle term conceived only between two motions." [53]

With the works of Mach, Kirchhoff, and Hertz the logical development of the process of eliminating the concept of force from mechanics was completed. This development in mathematical physics from the time of Newton onward was essentially an attempt to explain physical phenomena in terms of mass points and their spatial relations. Since the time of Keill and Berkeley, it became increasingly clear that the concept of force, if divested of all its extrascientific connotations, reveals itself as an empty scheme, a pure relational or mathematical function. Mach's, Kirchhoff's, and Hertz's contribution was the final stage in stripping off all the artificial trappings and embroideries from the concept; it was a process of purification, of methodological clarification. The concept of force in its metaphysical sense as causal transeunt activity had no place in the science of the

[53] *Ibid.*, p. 28.

empirically measurable. Whether the notion of force was tenable in general was outside the jurisdiction of physics.

This point was not always clearly understood in the nineteenth century. To safeguard "force" from such methodological attacks it was hypostatized as an incorporeal substance. Thus Spiller, for example, in his treatise *Der Weltäther als kosmische Kraft,* speaks of an independent substantiality of force. "No material constituent of a body, no atom, is in itself originally endowed with force, but every such atom is absolutely dead, and without any inherent power to act at a distance." [54] As a correlate to dead matter, he declares, force has to be conceived as an all-pervading "quasi-material presence" residing in the omniferous all-permeating ether.

A more serious argument in the defense of the concept of force, both in philosophy and in science, was the appeal to psychology and physiology. Hume's immediate successor, Thomas Reid, derived the idea of force, power or efficient cause from the consciousness we have of the operations of our own minds. "It is very probable that the very conception or idea of active power, and of efficient causes, is derived from our voluntary exertions in producing effects; and that, if we were not conscious of such exertion, we should have no conception at all of a cause, or of active power, and consequently no conviction of the necessity of a cause of every change which we observe in nature." [55] Although force or power is not itself an operation of the mind, nor an object of sense or of consciousness, it is associated with consciousness, since "we have very early, from our constitution a conviction or belief of some degree of active power in ourselves." [56] Every volition, according to Reid, implies a conviction of power.

A somewhat similar view was embraced by Kant's younger contemporary, François Pierre Gonthier Maine de Biran, whose

[54] Philip Spiller, *Der Weltäther als kosmische Kraft* (Berlin, 1873), p. 4.
[55] Thomas Reid, *Works,* ed. William Hamilton (Edinburgh, 1846), p. 604.
[56] *Ibid.,* p. 513.

personalistic philosophy has many points in common with the empirical intuitionalism of the chief of the Scottish school. The starting point of all science, declares de Biran, is our own direct experience of ourselves as causal agents. Thus the prototype of the idea of force is found in our own "will." Now, how are we aware that a certain action is not an involuntary act but the outcome of our ego as a source of forces? De Biran finds the criterion in the opposition of the muscles that is felt in the case of a voluntary activity. Although the consciousness of volitional force originates only through the kinesthetic sensation accompanying muscular contraction, volitional forces act on the (efferent) nerves stimulating this contraction. It is from the double character of the ego, being both an individual force and, on the other hand, inseparably united to a resisting organism, that we acquire the universal and necessary notion of force. It is not an innate notion, since it depends on a prior activity of the will, yet it is totally different from those general ideas produced by external observation to which Hume's criticism would apply.

Psychologico-physiological considerations such as these were adduced to save the concept of force from the destructive Humean criticism and to maintain the logical justification for employing it as a fundamental notion for the explanation of natural phenomena. In fact, it was claimed that the concept of force, manifested in our will, is the very first ascertained and infallible notion, prior even to the notion of our own existence. The Cartesian "Cogito ergo sum," according to de Biran, has to be replaced by "Volo ergo sum."

Arthur Schopenhauer adopted a similar view, although he opposed de Biran in many details. De Biran's view that the origin of the law of causality is found in the fact that the act of our will as cause is followed by the movement of our limbs as effect, is rejected by Schopenhauer. "We certainly do not recognize," he states in *The world as will and idea*, "the really immediate act of will as something different from the action of the body, and the two as connected by the bond of causality;

but both are one and indivisible. Between them there is no succession; they are simultaneous. They are one and the same thing apprehended in a double manner." [57] The concept of will, and consequently also the concept of force, are of all possible concepts the only ones that have their source not in the phenomenal, but come from within, proceeding from the most immediate consciousness, in which each of us knows his own individuality. The notion of force is thus, according to Schopenhauer, the last one to be questioned.

Such reflections on the immediacy of the psychological experience of forces were the bulwark from which the attack was launched against those who affirm that we know nothing but matter and motion. Sir John Herschel, for example, in his *Treatise on astronomy* wrote:

> Whatever attempts have been made by metaphysical writers to reason away the connection of cause and effect, and fritter it down into the unsatisfactory relation of habitual sequence, it is certain that the conception of some more real and intimate connection is quite as strongly impressed upon the human mind as that of the existence of an external world, the vindication of whose reality has, strange to say, been regarded as an achievement of no common merit in the annals of this branch of philosophy. It is our own immediate consciousness of effort, when we exert force to put matter in motion or to oppose and neutralize force, which gives us this internal conviction of power and causation, so far as it relates to the material world.[58]

Or let us quote William B. Carpenter, who, in his paper "The force behind nature," declared emphatically:

> Should we not think it absurd on the part of any one who possesses in the use of his hands the means of detecting the error of his visual perceptions, if he were to base a superstructure of reasoning — still more to found a whole system of philosophy — upon the latter alone? Yet such

[57] Arthur Schopenhauer, *Die Welt als Wille und Vorstellung* (1819); *The world as will and idea,* trans. R. B. Haldane and J. Kemp (London, seventh ed., 1909), vol. 2, p. 206.

[58] Sir John Herschel, "Treatise on astronomy," *Lardner's Cabinet Cyclopaedia* (London, 1830–1832), vol. 3, p. 232.

appears to me to be the position of those who deny our direct cognition of force.[59]

Indeed, it seems a persuasive argument against the elimination of "force" to claim that the concept of force stands in the same relation to the sensation of muscular effort as the concept of motion to the sensation of visual perception. The elimination of the concept of force in physical explanation, it was contended, is ultimately merely the replacement of muscular explanation of phenomena by an optical explanation. But why is it justifiable to prefer one sense to another? On the contrary, if one kind of sensation is to be preferred to the others, it should certainly be muscular sensation as being nearest to the psychological experience of volition, as shown in the preceding sections. The concept of force is "one of those universal ideas which belong of necessity to the intellectual furniture of every human mind."

Strong as these arguments seemed to be, it did not take long to invalidate them. William James, in his article "The Feeling of Effort," opposes the traditional theories on muscular exertion and its accompanying nerve-process, as advanced by Johannes Mueller, Bain, Jackson and Wundt, rejects the so-called feeling of innervation[60] and maintains "that the feeling of muscular energy put forth is a complex afferent sensation coming from the tense muscles, the strained ligaments, squeezed joints, fixed chest, closed glottis, contracted brow, clenched jaws, etc., etc."[61] James rejects the usual assertion that the resistance to our muscular effort is the only sense that brings us into close contact with reality independent of ourselves, and that reality, thus experienced, reveals itself in the form of a force like the force of effort which we exert ourselves.

[59] William B. Carpenter, "The force behind nature," *Popular Science Monthly 16*, 620 (1880).

[60] See, for example, Wilhelm Max Wundt, *Beiträge zur Theorie der Sinneswahrnehmung* (Leipzig and Heidelberg, 1862), p. 420.

[61] William James, "The feeling of effort," in *Collected essays and reviews* (New York, 1920), p. 154, where the article is reprinted from the *Anniversary Memoirs* of the Boston Society of Natural History (Boston, 1880).

The "muscular sense" being a sum of afferent feelings is no more a "force-sense" than any other sense. It reveals to us hardness and pressure as they do colour, taste, smell, sonority, and the other attributes of the phenomenal world. To the naïve consciousness all these attributes are equally objective. To the critical all are equally subjective. The physicist knows nothing whatever of force in a non-phenomenal sense. Force is for him only a generic name for all those things which will cause motion. A falling stone, a magnet, a cylinder of steam, a man, just as they appear to sense, are forces. There is no supersensible force in or behind them. Their force is just their sensible pull or push, if we take them naturally, and just their positions and motions if we take them scientifically. If we aspire to strip off from Nature all anthropomorphic qualities, there is none we should get rid of quicker than its "Force." How illusory our spontaneous notions of force grow when projected into the outer world becomes evident as soon as we reflect upon the phenomenon of muscular contraction. In pure objective terms (i.e., terms of position and motion), it is the relaxed state of the muscle which is the state of stress and tension. In the act of contraction, on the contrary, the tension is resolved, and disappears. Our feeling about it is just the other way, — which shows how little our feeling has to do with the matter.[62]

On the strength of considerations such as these it became more and more obvious that under no circumstances can "force" be identified with the feeling of strain, any more than there is any reason to identify "sound" or "colour" with the consciousness of sound or colour. The argumentations based on psychology and physiology, thus, were not so powerful as they seemed to be at first sight. That the experience of strain or tension is a purely subjective phenomenon and not necessarily a real property of physical nature, in the same sense as light, colour and sound are not necessarily outside the mind, was emphasized by Peter Guthrie Tait in his article "On force,"[63] which began, after a short lively controversy, a long series of publications in which philosophers of science in ever-increasing number expressed their favorable acceptance of the ideas advanced by Mach, Kirchhoff, and Hertz. The notion of force, says Tait, is a purely intellectual construction and "is no more an objective entity than say five per cent per annum is a sum of money." There is no point in listing all of

[62] *Ibid.*, p. 213.
[63] P. G. Tait, "On force," *Nature 17*, 459 (1869).

the authors, physicists, and philosophers who expressed similar ideas at that time; let us confine ourselves to some characteristic quotations. Tait's paper, just mentioned, was directed against the adoption of the concept of force as a necessarily fundamental notion in mechanics proper. However, this issue had already arisen, much earlier, with reference to the so-called vital forces. Criticizing the concept of vital force, DuBois-Reymond, in the preface to his *Untersuchungen über thierische Elektrizität,* discusses the concept of force in general. "Force," he says,

so far as it is conceived as the cause of motion, is nothing but an abstruse product of the irresistible tendency to personification which is impressed upon us; a rhetorical device, as it were, of our brain, which snatches at a figurative term, because it is destitute of any conception clear enough to be literally expressed . . . What do we gain by saying it is reciprocal attraction whereby two particles of matter approach each other? Not the shadow of an insight into the nature of the fact. But strangely enough, our inherent quest of causes is in a manner satisfied by the involuntary image tracing itself before our inner eye, of a hand which gently draws the inert matter to it, or of invisible tentacles with which the particles of matter clasp each other, try to draw each other close, and at last twine together into a knot.[64]

DuBois, in these words, rightly stresses the importance of analogy which was so instrumental in the formation of the concept of force in science. The irresistible tendency to personification and its role as a factor in the formation of our concept of force is also pointed out in a most lucid manner by Karl Pearson in his comprehensive *Grammar of science,* where he says:

Primitive people attribute all motion to some will behind the moving body; for their first conception of the cause of motion lies in their own will. Thus they consider the sun as carried round by a sun-god, the moon by a moon-god, while rivers flow, trees grow, and winds blow owing to the will of the various spirits which dwell within them. Slowly, scientific description replaces spiritualistic explanation. The idea, however, of enforcement, of some necessity in the order of a sequence, remains deeply rooted in men's mind, as a fossil from the spiritualistic explanation which

[64] Emil DuBois-Reymond, *Untersuchungen über thierische Elektrizität* (Berlin, 1848–1860), p. xi. Cf. Emil DuBois-Reymond, ***On animal electricity: being an abstract,*** ed. H. B. Jones (London, 1852).

sees in will the cause of motion. The notion of force as that which necessitates certain changes or sequences of motion, is a ghost of the old spiritualism.[65]

Since this phantom of force plays such a preëminent role in Newtonian mechanics, Pearson characterizes Newtonian dynamics as follows: "The Newtonian laws of motion form the starting-point of most modern treatises on dynamics, and it seems to me that physical science, thus started, resembles the mighty genius of an Arabian tale emerging amid metaphysical exhalations from the bottle in which for long centuries it has been corked down."[66]

Henri Poincaré discusses the anthropomorphic origin of the concept of force at some length in his *Science and hypothesis*. Anthropomorphism, he declares, has played an important role in the genesis of mechanics; it may, perhaps, even have an heuristic value also in the future by supplying symbolic notions that seem to be convenient to some minds. But it can never serve as the foundation of truly scientific reasoning.[67] Needless to say, Poincaré accepts Kirchhoff's definition of "force" (as the product of mass and acceleration), a definition that seems to him as most conforming to his conventionalistic point of view.

An interesting, but problematic, argument in favor of the ideas expressed by Mach, Kirchhoff, and Hertz was advanced at the beginning of the present century by Bertrand Russell, who bases his contention on his philosophy of mathematics. In his attempt to reduce pure mathematics, including geometry and even rational dynamics, to a theory dealing exclusively with concepts definable in terms of certain fundamental logical concepts (constants) and principles, Russell adopts Weierstrass's approach to the theory of real functions and the differential and integral calculus; Weierstrass developed analysis without the

[65] Karl Pearson, *The grammar of science* (1892; Dent, London, 1949), p. 104.

[66] *Ibid.*, p. 274.

[67] "L'anthropomorphisme a joué un rôle historique considérable dans la genèse de la mécanique . . . mais il ne peut rien fonder qui ait un caractère vraiment scientifique." Henri Poincaré, *La science et l'Hypothèse* (Paris, 1902), p. 136; in English, *The foundations of science*, trans. G. B. Halsted (Science Press, Lancaster, Pa., 1946), p. 104.

use of the concept of the infinitesimal. In consequence of such a view on the derivative of a function Russell concludes that the notion of a state of motion has to be entirely rejected.

Motion consists merely in the occupation of different places at different times, subject to continuity . . . There is no transition from place to place, no consecutive moment or consecutive position, no such thing as velocity except in the sense of a real number which is the limit of a certain set of quotients. The rejection of velocity and acceleration as physical facts (that is, as properties belonging at each instant to a moving point, and not merely real numbers expressing limits of certain ratios) involves some difficulties in the statement of the laws of motion; but the reform introduced by Weierstrass in the infinitesimal calculus has rendered this rejection imperative.[68]

From these premises Russell draws the conclusion that the notion of force ought not to be introduced into the principles of dynamics. The reason is this:

Force is the supposed cause of acceleration: many forces are supposed to concur in producing a resultant acceleration. Now an acceleration . . . is a mere mathematical fiction, a number, not a physical fact; and a component acceleration is doubly a fiction, for, like the component of any other vector sum, it is not part of the resultant, which alone could be supposed to exist. Hence a force, if it be a cause, is the cause of an effect which never takes place.[69]

Russell's denial of the validity of the concept of force hinges thus on two provocative arguments: first, a limit, a mere number, that is, the second derivative of distance with respect to time, cannot be a real (physical) event, and consequently it cannot be the effect of anything; second, components of accelerations or of forces are merely fictitious.

The present author, who has the highest respect for Russell's contributions to logic and metamathematics, still is of the opinion that Russell's argumentation here is not unimpeachable. It seems to be theoretically possible to recast theoretical physics in such a way that all differential equations would be replaced by

[68] Bertrand Russell, *The principles of mathematics* (1903; Norton, New York, 1943), p. 473.

[69] *Ibid.*, p. 474.

difference equations, a procedure that need not necessarily destroy the conceptual superstructure of mechanics, although it would lead to formidable technical difficulties. In fact, the possibility of some kind of "discrete" or discontinuous dynamics based on the "smallest length" or "fundamental length l_0," the hypothetical radius of the electron, has recently been considered quite seriously as a concept that meets certain difficulties in quantum electrodynamics.[70]

Russell's characterization of "force" as a fiction on the grounds of mathematical reasoning was not an important factor in support of the new descriptive mechanics. More effective for the general acceptance of these ideas was the approval expressed by modern philosophy and epistemology in particular. To mention only one example, let us quote a passage from Hans Vaihinger's *Philosophy of "as if"*:

> One of the most important fictions that arise through isolatory abstraction is that notorious and frequently dangerous product of the imagination, the concept of force . . . If two events, one preceding and the other following, are united by a constant bond, we call that peculiarity of the antecedent event, which consists in its being followed by another event, its "force," and measure this force in terms of the magnitude of its effect. In reality only sequences and coexistences exist, and we ascribe "forces" to things, by regarding the actual phenomena as already possible and then hypostasizing these possibilities and peculiarities, and separating them from the rest as real entities.[71]

It should not, however, be thought that this acceptance is unanimous among modern philosophers of science. Thus, for example, C. D. Broad in a detailed discussion on the concept of force states: "It seems clear to me that no one ever does mean or ever has meant by "force" rate of change of momentum." He goes on to say that undoubtedly the second law of motion,

[70] Cf. Werner Heisenberg, "Die Grenzen der Anwendbarkeit der bisherigen Quantentheorie," *Zeitschrift für Physik 110*, 251 (1938); Arthur March, *Natur und Erkenntnis* (Springer, Vienna, 1948); Arthur March, *Quantum mechanics of particles and wave fields* (Wiley, New York, 1951), sect. 69, p. 269.

[71] Hans Vaihinger, *The philosophy of "as if"* (1911), trans. C. K. Ogden (Routledge and Kegan Paul, London, 1949), p. 197.

as stated by Newton, was not intended for a definition of force, but was what he calls a substantial statement about it. "Unquestionably the sensational basis of the scientific concept of force is the feeling of strain that we experience when we drag a heavy body along," he declares, and continues:

I do not understand that this historical fact is denied by the upholders of the "descriptive" (or better, "definitional") theory. What they would probably say is that, in this sense, force is purely human and has no relevance to the laws of Mechanics. We cannot seriously suppose, e.g., that the sun feels a strain in keeping the earth in its orbit, as we do when we whirl a weight on a string. Hence it is argued that what we mean, when we say that the sun exerts a force on the earth, cannot be derived from the experiences of strain which we feel.

In answer to Broad's objection, so far, it should be said that no exponent of descriptive mechanics ever denied the fact that historically the scientific concept of force was formed in analogy to the sensation of strain or effort experienced by man; on the contrary, by calling traditional mechanics anthropomorphic they just emphasized this very fact. But let us see the main point of Broad's argument:

We must distinguish between our feeling of strain and the strain that we feel, just as we must distinguish between our feeling of movement and the movement which we feel ourselves to be making. Force is not supposed to be our feelings of strain; it is simply supposed that the strains which we feel are forces, or are indications of forces. It is of course absurd to suppose that the sun feels a strain when it pulls the earth; but this is absurd, not because the sun could not be subject to a strain, but because — having no mind — it cannot feel a strain or anything else. It is thus perfectly consistent for a man to describe forces as the sort of factors in nature which reveal themselves to us directly in our feelings of strain, and to add that inanimate bodies, like the sun, are subject to forces.[72]

Broad's argument, it seems, is nothing but a *petitio principii*. In his distinction between our feeling of strain and the strain felt, he interjects the notion of strain or force into physical

[72] C. D. Broad, *Scientific thought* (1923; Humanities Press, New York, 1952), p. 162.

reality. Properly speaking, he should differentiate between our feeling of strain and the physical phenomenon associated with our sensation of strain. Whether this objective phenomenon has any characteristics of what we call strain is a nonsensical question for physics. And how to describe this phenomenon is a matter of methodological convenience.

As a last reference in our survey of opinions pro and contra the necessity of the concept of force in mechanics let us quote an advocate of modern operationalism who, needless to say, rejects force as an entity independent of motion. Philipp Frank, in his article "Foundations of physics," declares emphatically: "The insistence upon the use of the term 'force' as an 'entity of its own' has its sources only in some psychological connotations of this word. We can quote the example of the Nazi philosophy in which the word 'force' is regarded as dear to the mind of the Nordic race. Every attempt to introduce a definition of 'force' by motion is branded as an act of the enemies of the Nordic race."[73]

In conclusion of the present chapter and in anticipation of the analysis of the concept of force in modern nonrelativistic mechanics as outlined in the following chapter, it should be kept in mind that generally throughout these considerations an *inertial system* is presupposed. Thus, for example, Mach's definitions of mass and force refer to an inertial system without raising the question whether the assumption of such a system does not presuppose the concept of force and does not consequently lead to a vicious circle. A rigorous solution of this intricate problem from the logical point of view is possible only within the framework of the theory of relativity.

[73] Philipp Frank, "Foundations of physics," *International encyclopedia of unified science* (University of Chicago Press, Chicago, 1955), vol. 1, p. 445.

CHAPTER 12

THE CONCEPT OF FORCE IN CONTEMPORARY SCIENCE

With the works of Mach, Kirchhoff, and Hertz the process of eliminating the concept of force from mechanics completed its logical development. Let us pause at this point for a moment and view this process as a whole in retrospect. At first, Kepler found in the notion of force (*vis*) a convenient concept to connect changes of speed in planetary motion with changes of distance. As we have shown in Chapter 5, it was essentially a methodological device that induced him to introduce this concept into mechanics — although he himself was unaware of the methodological aspect involved. We have also seen, in the chapters preceding Chapter 5, how the concept of force during its prescientific and semiscientific stages became loaded with a multitude of metaphysical, spiritual, and other extrascientific connotations. These connotations and associations formed an impressive psychological background, and the concept of force, when viewed against this background, seemed to be a suitable logical instrument to satisfy the human desire for causal explanation.

With the rise of Newtonian dynamics and its interpretation along the lines of Boscovich, Kant, and Spencer, the concept of force rose almost to the status of an almighty potentate of totalitarian rule over the phenomena. And yet, since the very beginning of its early rise to power, revolutionary forces were at work

(Keill, Berkeley, Maupertuis, Hume, d'Alembert) which in due time led to its dethronement (Mach, Kirchhoff, Hertz). This movement in mathematical physics, from the time of Newton onward, was essentially an attempt to explain physical phenomena in terms of mass points and their spatial relations. For it became increasingly clear that the concept of force, if divested of all its extrascientific connotations, reveals itself as an empty scheme, a pure relation. In fact, like "the king for a day" in the fairy tale, it came back to where it started. Hertz's conception of force "as a middle term between two motions" was, in fact, Kepler's starting point. True, it must be admitted that Kepler was not so conscious of what he did as Hertz was of what he undid. Yet was that the only difference? Was the whole story of the concept of force in classical mechanics merely the creation of an illusion and its evanescent dissolution? Our answer is: No! The history of physics shows clearly that the introduction of the concept of force led to a methodological unification of the conceptual scheme of science. Since the *raison d'être* of a scientific concept and its importance lie in the methodological function it performs, the concept of force in classical physics was not merely a will-o'-the-wisp. On the contrary, the concept of force played a most constructive role in the advancement of science and therefore wholly justified its existence.

"Force," so to say, was the common denominator of all physical phenomena and seemed thereby to be a promising instrument to reduce all physical events to one fundamental law. In fact, Newton's doctrine of forces already gave rise to such "unifying theories." Let us mention at present only two examples: one in Gowin Knight's *Attempt to demonstrate, that all the Phaenomena in Nature may be explained by two simple active Principles, Attraction and Repulsion*[1] and the other in Torbern Bergmann's

[1] Gowin Knight, *An attempt to demonstrate, that all the phaenomena in nature may be explained by two simple active principles, attraction and repulsion, wherein the attractions of cohesion, gravity and magnetism are shewn to be one and the same* (London, 1748).

Dissertation on elective attraction,[2] in which chemical affinities are also reduced to Newtonian attraction. Insufficient and unsuccessful as these attempts were, they are nevertheless a clear expression of the human endeavor to bring the whole of nature under the compass of one single law. But more important, perhaps, is that the concept of force was instrumental in the construction of the concept of energy, a notion whose contribution to a unified conception of physical phenomena is unquestioned.

It may be contended, and in fact has been contended, that the concept of potential energy is not less mystical than the concept of force. Hertz, J. J. Thomson, and others at the end of the nineteenth century anticipated this objection and attempted indeed to reduce the allegedly confused notion of potential energy, again with the aid of concealed masses and their motions, to the concept of kinetic energy, complementary to the kinetic energy of visible masses. The visible system of masses derives from the concealed masses just the increase of kinetic energy that it may show, and delivers to them just the amount of kinetic energy that it seems to lose. Although all these attempts have the advantage, as Thomson declared, "of keeping before us the idea that it is one of the objects of Physical Science to explain natural phenomena by means of the properties of matter in motion," the real motive behind these reductions are the methodological advantages implied. "Some of the theorems in dynamics become very much simpler from this point of view . . . The Principle of Least Action takes the very simple form, that with a given quantity of energy any material system will by its unguided motion go along the path which will take it from one configuration to another in the least possible time." [3]

The main advantage, however, of the concept of force — and this brings us to the status of our concept in present-day physics

[2] Torbern Bergmann, *A dissertation on elective attraction* (London and Edinburgh, 1785).

[3] J. J. Thomson, *Applications of dynamics to physics and chemistry* (London and Cambridge, 1888), pp. 14, 15.

— is that it enables us to discuss the general laws of motions irrespective of the particular physical situation with which these motions are associated. The concept of force in contemporary physics plays the role of a methodological intermediate comparable to the so-called middle term in the traditional syllogism. In order to show that "Socrates is mortal," we introduce the middle term "man" and state the two premises: (1) All men are mortal; (2) Socrates is a man. In our final conclusion, "Socrates is mortal," the middle term "man" drops out. Likewise, to show or to predict that a certain body *A* moves on a certain trajectory *B*, when surrounded by a given constellation of bodies *C*, *D*, . . . , which may be gravitating, electrically charged, magnetized, and so forth, we introduce the middle term "force" and state the two "premises": (1) The constellation *C*, *D*, . . . gives rise to a force *F*; (2) the force *F* (according to the laws of motion) makes the body *A* move on the trajectory *B*. In our final conclusion, "Body *A*, surrounded by *C*, *D*, . . . under the given circumstances, moves along trajectory *B*," the middle term "force" again drops out. Instead of connecting directly the kinematic behavior of body *A* with the arbitrarily given configuration *C*, *D*, . . . , we are splitting the situation up, so to say, into two parts. This methodological device enables us to study the kinematic aspects prior to, and independent of, the particular physical situation of the bodies concerned. Within the framework of classical mechanics, that is, so long as we confine ourselves to ordinary velocities, experience now shows that the product of the inertial mass m of our test body *A* and its acceleration a, that is, ma, is a function Φ of the total configuration under discussion. Configuration here means the amounts of the gravitational masses, electric charges, magnetic moments, and so forth, as well as the geometric constellations involved.[4] For the sake of brevity, let us denote this configuration by the symbol X. Thus

[4] In general, as, for example, in electrodynamic theory, the relative states of motion (of the charges) under discussion may also form part of the configuration. Clearly the mass, charge, polarization, and so forth, of the test body itself form part of this configuration.

$ma = \Phi(X)$. Mass m and acceleration a, taken separately, are not functions of X. Now, if we replace our test body of mass m with another body of mass m', the latter body, *ceteris paribus*, will move with an acceleration a' that satisfies the equation $ma = m'a'$. Whatever test body we insert, the product of its inertial mass by the acceleration is a single-valued function of the configuration X. The constancy of this product with respect to the various test bodies suggests our giving it a name of its own: we call it "force." Such a nominal definition (force = mass × acceleration) is of course an analytic statement or a tautology, as every nominal definition is. These remarks, it is hoped, will clarify the much-disputed logical status of Newton's second law: it is an empirical statement with respect to the fact that just the product "mass times acceleration" is a single-valued function of the configuration X. It is an analytic statement with respect to the nominal definition of force. Furthermore, if we postulate that the product ma, being a function of X, vanishes at infinity, or, in other words, if we postulate that the function Φ tends to zero if the distance variables in its argument increase indefinitely, we express the law of inertia. Newton's first law may thus be conceived as a boundary condition of the function Φ.

The replacement of the concept of force in classical physics by the concept of a functional dependence raises, however, an interesting question. In our expression $ma = \Phi(X)$, m signifies the inertial mass of the accelerated body. It was tacitly assumed that m can be defined and determined without any reference to the concept of force. Mach's definition and determination of mass as the negative inverse ratio of the mutually induced accelerations, although based on the assumption of "induced" accelerations, can be viewed as a purely kinematic procedure and seems to satisfy our quest for such a definition and determination. A more profound analysis of Mach's determination of mass, however, will show that the situation is not this simple. Mach's definition and determination, as originally conceived (see p. 221), presupposes that the two bodies under discussion form an isolated system. The question may now be asked whether Mach's approach holds also in the more general case of n bodies forming an isolated system if $n > 2$. It can easily be shown that, if $n \geqslant 5$, Mach's procedure breaks down, as it does for $n = 4$ if the accelerations and motions of the bodies under discussion are coplanar, and even for $n = 3$ if they are collinear.

In order to prove this theorem, let $\mathbf{a}_k$ be the (vector) acceleration of the kth body ($k = 1,2,\ldots,n$) due to the presence of the remaining $n - 1$ bodies of the isolated system. In view of Mach's Third Experimental Proposition (see p. 221), according to which the accelerations induced are independent of each other, $\mathbf{a}_k$ is the resultant of the individual $n - 1$ accelerations $\mathbf{a}_{ki}$, where $\mathbf{a}_{ki}$ denotes the acceleration of the kth particle due to its interaction with the ith particle alone. Let $\mathbf{u}_{ki}$ represent the unit vector in the direction from the kth particle to the ith particle. According to Mach's First Experimental Proposition, $\mathbf{a}_{ki}$ has the direction of $\mathbf{u}_{ki}$, that is, $\mathbf{a}_{ki} = \alpha_{ki}\mathbf{u}_{ki}$.

Hence $$\mathbf{a}_k = \sum_{i=1}^{n} \alpha_{ki}\mathbf{u}_{ki}. \qquad (k = 1,2,\ldots,n;\ \alpha_{kk} = 0).$$

The $\mathbf{a}_k$ and $\mathbf{u}_{ki}$ are determinable by observation and measurement. Thus we have n vector equations in three-dimensional space, or $3n$ algebraic linear equations for the $n(n - 1)$ unknowns α_{ki}. Consequently, the α_{ki} are determinate only if $n(n - 1) \leqslant 3n$, that is, if $n - 1 \leqslant 3$ or $n \leqslant 4$. Since the α_{ki} determine the relative masses or mass ratios, Mach's method becomes inoperative as soon as $n \geqslant 5$.

So far we have assumed that the values of $\mathbf{a}_k$ and $\mathbf{u}_{ki}$, which of course depend upon the time t, are taken with respect to one common value of the variable t. However, the proof can easily be generalized for nonsimultaneous observations and measurements. If these are performed at s different instants t_λ ($\lambda = 1,2\ldots,s$), we obtain

$$\mathbf{a}_k(t_\lambda) = \sum_{i=1}^{n} \alpha_{ki}(t_\lambda)\mathbf{u}_{ki}(t_\lambda). \qquad (k = 1,2,\ldots,n;\ \lambda = 1,2,\ldots,s)$$

These are $3ns$ linear equations for $sn(n-1)$ unknowns and they lead to solutions only if $n \leqslant 4$, as before.

The generalization of Mach's procedure for the determination of masses in the case of an isolated system of n bodies, if $n > 4$, requires, as we see, an additional assumption. Thus, for instance, Newton's law of gravitation may be used, giving the equations

$$m_k \mathbf{a}_k(t_\lambda) = G \sum_{i=1}^{n} \frac{m_i m_k}{r_{ik}^2} \mathbf{u}_{ik}(t_\lambda), \qquad \begin{matrix}(k = 1,2, \ldots, n; \\ \lambda = 1,2, \ldots, s)\end{matrix}$$

where G is the constant of universal gravitation and r_{ik} the distance between the ith and the kth bodies. Since the $\mathbf{a}_k$, $\mathbf{u}_{ik}$, and r_{ik} are observable, the m_k are determinable, provided s is sufficiently large.

The reference to any law of force, however, is not unavoidable. It can easily be shown that the law of conservation of linear momentum, for example, is wholly sufficient to supply the equations necessary for a unique determination of the masses of the system under consideration. Needless to say, so far our discussion has been confined to classical mechanics, in which the mass of a particle is independent of the velocity within a given inertial system.

In conclusion of our discussion of the concept of force in the modern treatment of classical mechanics one further point needs clarification. It may be argued that the notion of force, as proposed in the last sections, may well apply to dynamics proper, but not to statics, since in statics no motions and consequently no accelerations are involved. This objection was anticipated by Mach. "Force," he says, "is any circumstance of which the consequence is motion. Several circumstances of this kind, however, each single one of which determines motion, may be so conjoined that in the result there shall be no motion. Now statics investigates what this mode of conjunction, in general terms, is. Statics does not further concern itself about the particular character of the motion conditioned by the forces."[5] Statics thus viewed is a particular branch of dynamics, namely, the theory of

[5] Ernst Mach, *The science of mechanics,* trans. T. J. McCormack (Open Court, La Salle, Ill., 1942), p. 95.

the equilibrium of forces. Thus, the conception of force as outlined previously can be carried over to statics as well.

Modern physics recognizes the concept of force in both statics and dynamics, and thereby in every other field of physics as far as motive forces are concerned, as a methodological intermediate that in itself carries no explanatory power whatever. It is a construct to which no immediate rule of interpretation or epistemic correlation can be attached; only a long chain of logical relations, leading over the concepts of mass and acceleration, themselves complicated constructs, refers it to the data of sensory experience. In recognition of the highly abstract character of the concept of force, modern physics saw some justification in modifying its logical status and considering it as based on the concept of work or energy. In fact, it is often claimed that the concept of work or energy has a more decisive and fundamental role in present-day physics than that of force, and in usurping the concept of force by energy, only more justice is done to the physical nature of force. In consequence of considerations such as these, force is defined as the space rate of change of energy: The vectorial components f_n of a force in an n-dimensional coördinate system mapped out by the vectors $x_1, x_2, \ldots, x_n$ are those numbers f_n which determine how much work is performed during each of the virtual displacements x_n of its point of application. If an arbitrary displacement $r = r^1x_1 + r^2x_2 + \ldots + r^nx_n$ is given, the work performed is equal to $\sum_{i=1}^{n} f_i r^i$. In short, force is a covariant vector $(f_1, f_2, \ldots, f_n)$ satisfying the equation

$$\text{work} = \sum_{i=1}^{n} f_i r^i,$$

or an equivalent equation involving the use of the integral.

This definition corresponds to the more abstract treatment of these concepts in modern theoretical mechanics. Here (ordinary) forces may be considered as particular cases of the so-called "generalized forces Q_k." If in Lagrange's equations of the second kind the coördinates q_k, in the expressions $Q_k \delta q_k$ of the right-hand side, have the meaning of length, the coefficients Q_k define the components of an ordinary force.

The equation $ma = \Phi(X)$, with the boundary condition that Φ vanishes at infinity, is, as we have seen, the modern version of Newton's first two laws of motion. The classical treatment of the gravitational two-body problem is a simple example. Let m be the mass of the test body and M that of the other body and let r be the distance between the two; the function Φ is then given by the well-known expression

$$\Phi = G\frac{mM}{r^2}.$$

If m and M are measured in grams and r in centimeters, the gravitational constant G is equal to 6.67×10^{-8} dyne cm^2/gm^2. The inertial masses m and M, when regarded as constituents of the configuration X, are spoken of as "gravitational masses." Substituting this expression for the function Φ in our law of motion, we obtain

$$a = G\frac{M}{r^2},$$

and infer, for example, that all bodies on the surface of the earth fall with the same acceleration.

For classical mechanics the statement $ma = \Phi(X)$ is the representation of an ultimately irreducible principle. Any attempt of a mechanistic theory to replace, in the case of gravitation, this functional concept of force by the assumption of a chain of causal contiguous actions (Chapter 10), was doomed to failure. In the case of electric forces the situation is not much different, in spite of the electric field theory of Faraday and Maxwell, certainly the most profound transformation experienced by the foundations of physics between the time of Newton and that of Einstein. True, the idea of stresses in the ether seemed at first to be a constructive step forward in the attempt to find a model that could substitute for the purely mathematical character of the functional relation Φ. Yet, as is well known and need not be elaborated in detail, Maxwell's mechanical ether models and, moreover, the conception of an all-pervading ether itself, proved untenable.

The conception of force as a purely functional relation is in full accord with its application in quantum mechanics and nuclear physics. If in classical physics the use of the concept of force is essentially a device for economy of thought, based on analogy with human experience, it is more so in quantum mechanics. Whether quantum mechanics is formulated as an operator calculus in which dynamical variables, such as coördinates or momentum components, are represented by matrices, or whether the Schrödinger formalism is adopted, the concept of force, via the notion of potential energy, is introduced in complete analogy to macroscopic dynamics and is consequently, strictly speaking, an analogy of an analogy. No matter which mode of representing the theory is adopted, a certain formal analogy relation has to be established between the equations of classical dynamics for a given system of particles on the one hand and the quantum formalism on the other. In fact, analogy relations, such as the correlation between the classical Hamiltonian

$$H(p,q) = T(p) + V(q) = \frac{p^2}{2m} + V(q)$$

and the corresponding operator expression or matrix expression in which the vector differential operator $(h/2\pi i)$grad replaces p, seem to be indispensable for the development of the theory. They not only guide us in translating the equations of classical mechanics for a given atomic or molecular system into quantum-mechanical language, but also assist us in interpreting the outcome of our quantum-mechanical calculations in a meaningful sense. The wave equation or matrix expression for a system involving six coördinates and the function e^2/r owes its meaning as well as its very existence to the classical description of a system of two particles with inverse-square attraction.

No one has ever directly demonstrated the force of attraction between, say, a proton and an electron. And yet, in writing Schrödinger's equation for such a system, we use the term e^2/r for the potential energy, carrying it over, so to say, from classical dynamics as a generalization ultimately based on the concept of force. It is of course the correspondence between the mathematical result and the experimental result that justifies this procedure. Even according to Dirac's theory of the electron, which is perhaps the most satisfactory one we have so far, all that can be stated is: the electron behaves *as if* it is attracted or repelled, *as if* it has an internal angular momentum and an associated magnetic moment.

As long as quantum mechanics has to borrow part of its basic conceptions from classical dynamics, as long as it has no logically and methodologically independent conceptual apparatus, it cannot be expected to lead to a revision of the classical conception of force. This, of course, does not mean that it cannot be successful in its new interpretation of certain macroscopic forces, such as elasticity or magnetism. These contributions, however, have clearly no import on the logical, epistemological, or methodological status of the concept of force.

An excellent illustration of these contentions is the present state of the theory of nuclear forces, incomplete and problematic as it is. Experience (scattering experiments) shows clearly that nuclear physics is confronted with a new kind of (short-range) force, wholly different from gravitational or electromagnetic forces. And yet, lacking a satisfactory fundamental theory of nuclear forces, modern physics attempts as far as possible to follow the conventional approach and to model its conceptual scheme in analogy to that adopted in the study of classical forces.

As soon, however, as analogy considerations lead to conclusions inconsistent with experience or unsuccessful in the theoretical interpretation of experiments, nuclear theory has to explore new paths and unconventional methods. Thus, for example, nuclear physics tries to adhere to the traditional assumption that the forces involved in nuclear processes are similar in their general character to "atomic forces," that is, that they are "two-body forces" acting between pairs of nucleons while the presence of other nucleons does not modify the force between the two given nucleons (in analogy to the Coulomb law of force according to which the interaction between two charges is not affected by the presence of a third charge). There is, however, a growing body of opinion that part of the forces in the nuclear interior are "many-body forces," that is, interactions that do become modified by the presence of other particles, and that some of these forces are tensor forces (noncentral, spin-dependent forces).

Once particle interactions are interpreted by means of an intervening field mechanism (in analogy to quantum electromagnetic theory), it becomes obvious that the traditional description of the dynamical behavior of the particles, based on action-at-a-distance forces (or potentials), leads only to approximate solutions. The traditional representation of forces by two-body interactions in which the potential of a system of n particles is equal to the sum of the individual two-body potentials,

$$V(r_{12}, r_{13}, \ldots) = \sum_{i,j} V_{ij}\,(r_{ij}),$$

is consequently regarded only as a first approximation which proves to be excellent for the description of electrons in atomic systems but inadequate for the treatment of heavy particles in the nucleus.[6] If the force between any two particles is regarded as dependent also upon the positions of other particles, many-body forces can be represented by means of many-body potentials — in a relatively simple case, for example, by three-body potentials:

$$V(r_{12}, r_{13}, \ldots) = \sum_{i,j,k} V_{ijk}\,(r_{ij}, r_{ik}, r_{jk}).$$

Here the potential between the ith and the kth particles does not

[6] H. Primakoff and T. Holstein, "Many-body interactions in atomic and nuclear systems," *Physical Review 55*, 1218 (1939).

depend only upon their mutual separation r_{ij} but is also a function of their respective distances from the *k*th particle.

It must be recognized, of course, that the conception of force as a purely functional relation is not confined to the notion of "two-body forces." Although in classical physics the accelerative action of pairs of particles, in general, determines the behavior of a many-particle system, the assumption of a more general type of force, such as the above-mentioned three-body forces, does not necessarily lead to inconsistencies. The introduction of such forces would only unnecessarily complicate the mathematical apparatus involved. For this reason theories of this type of force have rarely been developed. Also progress in this direction was probably precluded, as Henry Margenau has pointed out,[7] by the historical accident that Hermann Helmholtz based his demonstration of conservation of energy on the assumption of two-body (central) forces.[8]

Nuclear theory regards the characteristic properties of nuclear forces (attraction, repulsion) as depending upon the "state" of the two nucleons with respect to each other, that is, as a function of their configuration. In order to account for the saturation of binding energy and of nuclear density Werner Heisenberg[9] and Ettore Majorana[10] introduced so-called "exchange forces." This was done in analogy to the quantum-mechanical theory of covalent bonds, such as exist between two hydrogen atoms in the hydrogen molecule: the chemical force is attractive if the wave function is symmetric under exchange of the coördinates of the electrons and is repulsive if the wave function is antisymmetric in this respect. Since the Pauli exclusion principle precludes the possibility of having only symmetric pairs in a nuclear wave function, the above-

[7] Henry Margenau, "The exclusion principle and its philosophical importance," *Philosophy of Science 11*, 205 (1944).

[8] Hermann Helmholtz, *Über die Erhaltung der Kraft* (1847), in Ostwald's *Klassiker der exacten Wissenschaften* (Engelmann, Leipzig, 1889), No. 1, p. 11.

[9] Werner Heisenberg, "Über den Bau der Atomkerne," *Zeitschrift fur Physik 77*, 1 (1932).

[10] Ettore Majorana, "Über die Kerntheorie," *Ibid. 82*, 137 (1933).

named saturation effects in nuclear physics could easily be accounted for, once "exchange forces" were admitted.

The notion of "exchange force," although a purely quantum-mechanical concept, is essentially not a new conception of force as such. The fact that an "exchange force" has no analogue in classical physics is not a consequence of a new conception of force. Its unconventional character lies rather in the assumption of a continuous exchange of particles which accompanies the interaction and transmits the force — a process that receives its operational justification through Heisenberg's uncertainty principle.[11]

Yukawa's meson theory, according to which the interaction between nucleons arises from the virtual exchange of mesons whose experimentally ascertainable mass indicates the range of the force, involves therefore no new aspects of the concept of force as such. Nor is a revision of our conception of force required by the classifications of nuclear forces into Wigner forces (short-range nonexchange forces), Heisenberg forces (exchange of both the position and spin coördinates of the two interacting nucleons), Majorana forces (exchange of position coördinates), Bartlett forces (exchange of spin coördinates), or other forces as mixtures of the foregoing, or Serber forces, and so on. They express merely different methods of introducing the dependence of the forces on the "state" of the particle. In short, they are different formulations of Φ.

Let us now examine the position of the concept of force within the framework of the theory of relativity. The mechanics of special relativity led to only minor revisions, primarily in the mathematical aspects, of the concept of force. Here, as in classical mechanics, force is defined as the rate of change of momentum,

$$\mathbf{f} = \frac{d(m\mathbf{u})}{dt},$$

[11] David L. Falkoff, "Exchange forces," *American Journal of Physics 18*, 30 (1950).

an equation that allows for a change in mass as well as for a change in velocity.[12] More specifically, the components of the relativistic force are given by the time rate of change of the momentum components,

$$f_n = \frac{d}{dt}\left(\frac{m_0\dot{x}_n}{(1 - u^2/c^2)^{\frac{1}{2}}}\right),$$

where m_0 is the rest mass, $\dot{x}_n$ the time derivative of the coördinate x_n with respect to the local time t, and u the velocity of the moving body, that is,

$$u^2 = \sum_n \dot{x}_n^2.$$

The relativistic definition of force differs from the Newtonian definition (the second law of motion) essentially in the velocity dependence of the mass:

$$m = \frac{m_0}{(1 - u^2/c^2)^{\frac{1}{2}}}.$$

This is just the result of an attempt to retain as much of the Newtonian characteristics of the concept of force as possible. The following analysis will clarify this point.

Newton's principle of action and reaction (the third law of motion) is classically equivalent to the principle of conservation of momentum. In this latter form the principle is retained in special relativity. Now, if in any four-dimensional coördinate system ($x_1, x_2, x_3, x_4 = t$) the proper time interval $d\tau$ is defined by the equation

$$c^2d\tau^2 = c^2dt^2 - dx_1^2 - dx_2^2 - dx_3^2 = c^2dt^2(1 - u^2/c^2),$$

[12] This definition, given by Einstein in his article "Das Relativitätsprinzip und die aus demselben gezogenen Folgerungen," *Jahrbuch der Radioaktivität 4,* 411 (1907), was first used by Gilbert N. Lewis in his paper "A revision of the fundamental laws of matter and energy," *Phil. Mag. 16,* 705 (1908). Richard G. Tolman saw additional support for this definition in the fact that it led to a derivation of the equation for the Lorentz force from Maxwell's electromagnetic-field equations. See Tolman's article, "Note on the derivation from the principle of relativity of the fifth fundamental equation of the Maxwell-Lorentz theory," *Phil. Mag. 21,* 296 (1911).

$d\tau$ is obviously invariant with respect to Lorentz transformations. Consequently, the four components of the "Minkowski velocity" **v**,

$$v_1 = \frac{dx_1}{d\tau}, \quad v_2 = \frac{dx_2}{d\tau}, \quad v_3 = \frac{dx_3}{d\tau}, \quad v_4 = \frac{dt}{d\tau}$$

transform like coördinate intervals. If, furthermore, the principle of conservation of momentum is applied — for example, to collisions of free particles — simple considerations show that the only possible expressions that correspond to the classical quantities $m_0 dx_n/dt$ are $m_0 dx_n/d\tau$. Thus, only if the momentum components are defined by

$$m_0 \frac{dx_n}{d\tau} = \frac{m_0 \dot{x}_n}{(1 - u^2/c^2)^{\frac{1}{2}}}$$

does conservation of momentum in one coördinate system imply that momentum in another Lorentz system is also conserved.

Another point of similarity between the Newtonian and the relativistic concept of force is worth mentioning. If we multiply the fourth component of the "Minkowski velocity" by m_0c^2, we obtain

$$m_0c^2 \frac{dt}{d\tau} = \frac{m_0c^2}{(1 - u^2/c^2)^{\frac{1}{2}}} = m_0c^2 \left(1 + \frac{1}{2}\frac{u^2}{c^2} + \cdots\right)$$

$$= m_0c^2 + \tfrac{1}{2}m_0u^2 + \cdots,$$

which shows that (apart from the rest energy m_0c^2) $m_0c^2(1 - u^2/c^2)^{-\frac{1}{2}}$ is the kinetic energy E of the moving body. If **P** denotes the momentum,

$$\mathbf{P} = \frac{m_0\mathbf{u}}{(1 - u^2/c^2)^{\frac{1}{2}}},$$

we have the well-known relation

$$E^2/c^2 - P^2 = m_0^2c^2.$$

In combination with the definition of force, $\mathbf{f} = d\mathbf{P}/dt$, we finally obtain

$$\frac{dE}{dt} = \mathbf{f}\cdot\mathbf{u},$$

that is, the scalar product of force and velocity is equal to the time rate of change of work, which shows, again, the great analogy between the Newtonian concept of force and its counterpart in special relativity. However, in contrast to the Newtonian conception, it is easy to show that in relativity the quantity force, in general, is not codirectional with the acceleration it produces; one has only to carry out the differentiation, in a straightforward manner, in the defining formula of force. It is also easy to show that these force components have no simple transformation properties, although, as we have seen, their integration over a three-dimensional path corresponds to the change of the relativistic kinetic energy. If we carry out the differentiation of the momentum with respect to the proper time τ, instead of with respect to the system time t, we define what is generally called the "world force." All these modifications, important as they are from the mathematical point of view, do not radically affect the conception of force. Yet an important point should be noted: on grounds of the rejection of an absolute simultaneity of two distant events special relativity comes to the conclusion that action at a distance has to be excluded as a legitimate physical notion. Forces, in other words, can only be contact forces.

It was the theory of general relativity that led to a more profound revision of the concept of force, a process that is far from being completed even at present. In fact, it has been carried out successfully so far only with respect to gravitational forces. Its generalization for nongravitational forces, and primarily (and perhaps only) for electromagnetic forces, is intimately connected with the problem of the so-called "unified field theories."

In two respects the general theory of gravitation reduced the functional concept of force to more profound or fundamental relations. For the description of these modifications we shall, in general, refer to the example of gravitational forces, since in this case a consistent mathematical theory has been established while the incorporation of electromagnetic forces into a similar treatment is still a matter of dispute. Now, for classical physics, the fact that a certain uniquely determined product *ma* is associated with a given configuration Φ is, as we have seen, an ultimately irreducible datum. "Irreducible" and "unexplained" (that is, not subsumed under more general relations) are almost synonymous words. In a certain sense, therefore, the acceleration of an object, from the point of view of classical physics, may be regarded as a miracle. Apart from the purely mathematical relation there is no "bridge" that leads from "configuration" to "accelerated motion."

In contrast to classical physics, however, such a connecting link has been established in general relativity by Einstein's ingenious synthesis of geometry and gravitation.[13] Through the principle of equivalence, based on the proportionality of inertial and gravitational mass, gravitation reveals itself as merely an inertial force; it is fictitious to the same extent as a centrifugal force, for example. As explained in elementary mechanics, a revolving reference system of coördinates gives rise to centrifugal forces. Yet these forces disappear as soon as the motion is related to an inertial frame of coördinates; the inertial behavior of the body regulates its motion.

If we introduce a Riemannian coördinate system for the space-time continuum, x^1, x^2, x^3, x^4, the geometry of the continuum will be determined by the line element (that is, the four-dimensional distance between two neighboring events)

$$ds^2 = \Sigma g_{\mu\nu} dx^\mu dx^\nu, \text{ in which } g_{\mu\nu} = g_{\nu\mu}.$$

General relativity now assumes only one law of motion: a free particle moves on a geodesic line, defined by the equation

$$\delta \int ds = 0.$$

A "free" particle, here, means a particle not subjected to collisions or electromagnetic forces; it may, however, well be situated

[13] Albert Einstein, "Zur allgemeinen Relativitätstheorie," *Sitzungsberichte der preussische Akademie der Wissenschaften* (1915), pt. 2, pp. 778–786, 799–801.

in the proximity of other lumps of matter (presence of "gravitation").

Now an observer, unaware of the natural local geometry, may refer physical events in the external world that surrounds him to a Galilean frame of coördinates corresponding to the line element

$$ds'^2 = -(dx^2)^2 - (dx^2)^2 - (dx^3)^2 + (dx^4)^2, \quad (x^4 = ct)$$

this being the geometry with which he is most familiar (semi-Euclidean geometry). By experiment and observation our observer will soon realize that the particle will not follow the path described by the condition

$$\delta \int ds' = 0.$$

In his geometry this means that the particle will not move in a straight line with constant velocity. Its deviation from this path will be accounted for by his assumption of an agency that "causes" this discrepancy, in other words, by a "force." Gravitation, as manifested in the parabolic trajectory of a projectile, is a deviation of this kind. Thus, gravitational forces are the outcome of the application of a wrong metric. If the appropriate Riemannian metric had been used, in which the so-called gravitational potentials $g_{\mu\nu}$ are determined through the field equations by the mass-energy distribution, the trajectory of our projectile would identify itself as the geodetic line in four-dimensional space-time, corresponding to the given initial conditions. Whereas in classical mechanics the configuration X was the determinative element of motion, in general relativity it is the space-time continuum itself. Since the inertial behavior of a particle, namely, its motion on a geodetic line, is a "force-free" motion, and since in addition all free particles, in general relativity, move only on geodetic lines, providing the appropriate Riemannian metric has been applied, there is no need at all in general relativity for the concept of gravitational "force."

The elimination of the (relational) concept of force in general relativity is achieved, in principle, by the following methodological devices. "Force" is defined by the deviation of a particle from its "natural" path in space-time. The path of a particle in space-time is called "natural" if no forces are exerted on the particle. The vicious circle implied in these two definitions can be broken in two different ways: (1) a "natural" path in space-time may be interpreted as a uniform rectilinear motion with respect to the "inertial system of the fixed stars"; (2) a "natural" path in space-time may be interpreted as a geodesic in a Riemannian continuum. The former alternative, adopted by classical mechanics, leads to the existence of gravitational forces in combination with a Euclidean space-time; the second alternative, incorporated as a fundamental idea into the theory of general relativity, leads to a non-Euclidean space-time of variable curvature without gravitational forces. Only in the absence of (gravitating) matter does Riemannian space-time become Euclidean and geodetic lines degenerate into the straight lines of Euclidean geometry. Thus, gravitation in general relativity has not the character of a force. It is a property of space-time. Mechanical events are thus accounted for by purely geometrico-kinematic conceptions. Descartes's program has finally been carried out by Einstein! It is only natural, therefore, that general relativity does not include — at least not as a rigorous law — the principle of action and reaction nor its most important dynamical consequence concerning the motion of the center of mass.

Obviously, relativistic theories of gravitation, as proposed by Whitehead,[14] Birkhoff,[15] and others, in which the geometry of space-time is homaloidal (or Euclidean), do not possess the methodological simplification of eliminating the concept of force. Yet, even in Einstein's general relativity the miracle to which

[14] A. N. Whitehead, *The principle of relativity* (Cambridge University Press, Cambridge, 1922).

[15] G. D. Birkhoff, "Matter, electricity and gravitation in flat space-time," *Proceedings of the National Academy of Sciences 29*, 231 (1943).

we have alluded previously has not yet disappeared; it has been transferred, so to say, only to a different plane. It lies now in the functional relation between the space-time structure and the mass-energy distribution, or, in other words, in Einstein's field equations $R_{\mu\nu} - \frac{1}{2}g_{\mu\nu} R = -T_{\mu\nu}$ connecting the scalar Riemannian curvature R with the energy tensor $T_{\mu\nu}$ (in its phenomenological representation). Instead of $ma = \Phi(X)$, we have now the field equations as an ultimately irreducible principle. Once this assumption is made, the whole mechanics of gravitation consists merely in the solution of a single system of covariant partial differential equations.

For a consistent field theory the concept of a "particle" is extraneous. It seemed therefore very tempting to interpret mass points as singularities of the potentials of the field equations. Such a conception of matter as a singularity, moreover, is most attractive because it solves the problem of the spatial structure of particles in a most simple and radical way — by depriving them of any structure. William Kingdon Clifford's speculations on the identity of space and matter come thus to life again,[16] mathematically elaborated and modified in accordance with modern atomic theory. Such a theory would face serious difficulties from the viewpoint of physics (infinite total energy of the field, infinite electrostatic energy of point charge, and so on). The main difficulty from the viewpoint of mathematics is the excessive indeterminateness of the solutions of the field equations, if singularities are admitted. Under certain circumstances, however, some important restrictions on the singularities may arise, especially if the system of equations is to some extent interdependent. An elementary example is the constancy of charges in Maxwell's theory, a constancy that is a consequence of the equations of the electromagnetic field. In the theory of general relativity the identical vanishing of the divergence of the energy tensor of

[16] W. K. Clifford, *The common sense of the exact sciences*, ed. J. R. Newman (Knopf, New York, 1946), p. 202. Cf. Max Jammer, *Concepts of space* (Harvard University Press, Cambridge, 1954), pp. 160ff.

matter, corresponding to the conservation of impulse and energy, imposes certain conditions on the world line of the singularity, so that it cannot be prescribed arbitrarily. Let us describe the situation in more detail. Matter, it will be recalled, is represented in Einstein's equation $R_{\mu\nu} - \frac{1}{2}g_{\mu\nu} R = -T_{\mu\nu}$ by the local concentrations of the energy momentum tensor $T_{\mu\nu}$. A material particle will consequently be represented by a world tube of timelike direction and of small cross section. The tensor $T_{\mu\nu}$, which vanishes outside the world tube, is different from zero in its interior. Through a limiting process in which the world tube contracts, so to say, into a world line which now represents the path of our test particle (of negligible mass), the following theorem can be proved: the field equations, by virtue of Bianchi's identities, imply that the covariant divergence of the energy-momentum tensor vanishes everywhere. Finally, it can be shown that the source-free character of $T_{\mu\nu}$ restricts the world line of our particle to a geodesic.

Einstein and Grommer[17] showed in 1927 that the assumption of a static gravitational field, that is, a configuration of masses relatively at rest (in the absence of electromagnetic forces, of course), is incompatible with the field equations themselves. In other words, the motion of the mass particle is a consequence of the field equations. Similarly, Myron Mathisson,[18] in his first-approximation solution of the field equations for a moving charged particle in a gravitational field, has shown that the equations of motion reveal themselves as the conditions for the existence of solutions. In a paper entitled "The gravitational equations and the problem of motion,"[19] Einstein, Infeld, and Hoffmann have pointed out that the gravitational equations for

[17] A. Einstein and J. Grommer, "Allgemeine Relativitätstheorie und Bewegungsgesetz," *Sitzungsberichte der preussischen Akademie der Wissenschaften* (*Phys.-math. Klasse*), 1927, pp. 2–13, 235–245.

[18] Myron Mathisson, "Die Mechanik des Materieteilchens in der allgemeinen Relativitätstheorie," *Zeitschrift für Physik 67*, 826 (1931).

[19] A. Einstein, L. Infeld, and B. Hoffmann, "The gravitational equations and the problem of motion," *Annals of mathematics 39*, 65 (1938).

empty space are fully sufficient to determine the motion of matter, represented as singularities of the field. Among the nonlinear gravitational equations four differential relations can be established; the gravitational equations form an overdetermined system of equations. It is this overdetermination that is responsible for the existence of equations of motion. Thus, in contrast to classical electrodynamics in which Lorentz's force law is independent of Maxwell's field equations, in general relativity the law of motion is a necessary consequence of the field equations. The fact that the laws of motion in classical field theories are independent of the field equations is a consequence of the linearity of the latter. On the other hand, it is essentially the nonlinear character of the field equations in general relativity that made it possible to deduce the dynamical law from the field equations themselves. It must, however, be admitted that the approximation method employed by Einstein, Infeld, and Hoffmann applies only to the case of slowly varying fields. An attempt to attack this problem for arbitrarily strong fields was made in 1949 by Infeld and Schild.[20] Their derivation of the laws of motion is based on a particle picture of matter that avoids the introduction of an energy-momentum tensor altogether. Starting from the field equations $R_{\mu\nu} - \frac{1}{2}g_{\mu\nu} R = 0$, they consider a sequence of particles with masses m tending to zero and a corresponding sequence of metric fields $g_{\mu\nu}(x^\rho;m)$. These metric fields are expanded in powers of m and it is demonstrated that in the limit $m = 0$ the field equations identify the world line of the particle as a geodesic of the external "background" field.

These considerations have brought us to the brink of present-day research in theoretical physics. If it were possible to work out a unified field theory that subjects electromagnetic and possibly also nuclear forces to a similar treatment as gravitation,

[20] L. Infeld and A. Schild, "On the motion of test particles in general relativity," *Reviews of Modern Physics 21*, 408 (1949). See also A. Papapetrou, "Equations of motion in general relativity," *Proceedings of the Physical Society, London (A) 64*, 57 (1951).

it would lead us to a final stage in the history of the concept of force. While the modern treatment of classical mechanics still admitted, tolerantly, so to say, the concept of force as a methodological intermediate, the theory of fields would have to banish it even from this humble position.

INDEX

HARPER TORCHBOOKS / The Bollingen Library

Rachel Bespaloff — ON THE ILIAD. Introduction by Hermann Broch TB/2006

Elliott Coleman, *ed.* — LECTURES IN CRITICISM: *By R. P. Blackmur, B. Croce, Henri Peyre, John Crowe Ransom, Herbert Read, and Allen Tate* TB/2003

C. G. Jung — PSYCHOLOGICAL REFLECTIONS. Edited by Jolande Jacobi TB/2001

Erich Neumann — THE ORIGINS AND HISTORY OF CONSCIOUSNESS. *Vol. I,* TB/2007; *Vol. II,* TB/2008

St.-John Perse — SEAMARKS. Translated by Wallace Fowlie TB/2002

Jean Seznec — THE SURVIVAL OF THE PAGAN GODS: *The Mythological Tradition and Its Place in Renaissance Humanism and Art.* Illustrated TB/2004

Heinrich Zimmer — MYTHS AND SYMBOLS IN INDIAN ART AND CIVILIZATION TB/2005

HARPER TORCHBOOKS / The Academy Library

James Baird — ISHMAEL: *A Study of the Symbolic Mode in Primitivism* TB/1023

Herschel Baker — THE IMAGE OF MAN: *A Study of the Idea of Human Dignity in Classical Antiquity, the Middle Ages, and the Renaissance* TB/1047

Jacques Barzun — THE HOUSE OF INTELLECT TB/1051

W. J. Bate — FROM CLASSIC TO ROMANTIC: *Premises of Taste in Eighteenth Century England* TB/1036

Henri Bergson — TIME AND FREE WILL: *An Essay on the Immediate Data of Consciousness* TB/1021

H. J. Blackham — SIX EXISTENTIALIST THINKERS: *Kierkegaard, Jaspers, Nietzsche, Marcel, Heidegger, Sartre* TB/1002

Walter Bromberg — THE MIND OF MAN: *A History of Psychotherapy and Psychoanalysis* TB/1003

Abraham Cahan — THE RISE OF DAVID LEVINSKY. A novel. Introduction by John Higham TB/1028

Helen Cam — ENGLAND BEFORE ELIZABETH TB/1026

Joseph Charles — THE ORIGINS OF THE AMERICAN PARTY SYSTEM TB/1049

T. C. Cochran & William Miller — THE AGE OF ENTERPRISE: *A Social History of Industrial America* TB/1054

Norman Cohn — THE PURSUIT OF THE MILLENNIUM: *Revolutionary Messianism in Medieval and Reformation Europe and its Bearing on Modern Totalitarian Movements* TB/1037

G. G. Coulton — MEDIEVAL VILLAGE, MANOR, AND MONASTERY TB/1022

Wilfrid Desan — THE TRAGIC FINALE: *An Essay on the Philosophy of Jean-Paul Sartre* TB/1030

Cora Du Bois — THE PEOPLE OF ALOR: *A Social-Psychological Study of an East Indian Island. Vol. I,* 85 illus., TB/1042; *Vol. II,* TB/1043

George Eliot — DANIEL DERONDA. A novel. Introduction by F. R. Leavis TB/1039

John N. Figgis — POLITICAL THOUGHT FROM GERSON TO GROTIUS: 1414-1625: *Seven Studies.* Introduction by Garrett Mattingly TB/1032

Editors of *Fortune* — AMERICA IN THE SIXTIES: *The Economy and the Society* TB/1015

F. L. Ganshof — FEUDALISM TB/1058

G. P. Gooch — ENGLISH DEMOCRATIC IDEAS IN THE SEVENTEENTH CENTURY TB/1006

Francis J. Grund — ARISTOCRACY IN AMERICA: *A Study of Jacksonian Democracy* TB/1001

W. K. C. Guthrie — THE GREEK PHILOSOPHERS: *From Thales to Aristotle* TB/1008

Marcus Lee Hansen — THE ATLANTIC MIGRATION: 1607-1860 TB/1052

Alfred Harbage — AS THEY LIKED IT: *A Study of Shakespeare's Moral Artistry* TB/1035

J. M. Hussey — THE BYZANTINE WORLD TB/1057

Henry James — THE PRINCESS CASAMASSIMA. A novel. Intro. by Clinton Oliver. TB/1005

Henry James — RODERICK HUDSON. A novel. Introduction by Leon Edel TB/1016

Henry James — THE TRAGIC MUSE. A novel. Introduction by Leon Edel TB/1017

William James — PSYCHOLOGY: *The Briefer Course.* Edited with an Introduction by Gordon Allport TB/1034

Arnold Kettle — AN INTRODUCTION TO THE ENGLISH NOVEL. *Vol. I, Defoe to George Eliot,* TB/1011; *Vol. II, Henry James to the Present,* TB/1012

Samuel Noah Kramer — SUMERIAN MYTHOLOGY: *A Study of Spiritual and Literary Achievement in the Third Millennium B.C.* Illustrated TB/1055

Paul Oskar Kristeller — RENAISSANCE THOUGHT: *The Classic, Scholastic, and Humanist Strains* TB/1048

L. S. B. Leakey — ADAM'S ANCESTORS: *The Evolution of Man and His Culture.* Illustrated TB/1019

Bernard Lewis — THE ARABS IN HISTORY TB/1029

Ferdinand Lot — THE END OF THE ANCIENT WORLD AND THE BEGINNINGS OF THE MIDDLE AGES TB/1044

Arthur O. Lovejoy — THE GREAT CHAIN OF BEING: *A Study of the History of an Idea* TB/1009
Robert Lowie — PRIMITIVE SOCIETY. Introduction by Fred Eggan TB/1058
Niccolo Machiavelli — HISTORY OF FLORENCE AND OF THE AFFAIRS OF ITALY: *From the Earliest Times to the Death of Lorenzo the Magnificent.* Introduction by Felix Gilbert TB/1027
J. P. Mayer — ALEXIS DE TOCQUEVILLE: *A Biographical Study in Political Science* TB/1014
John U. Nef — CULTURAL FOUNDATIONS OF INDUSTRIAL CIVILIZATION TB/1024
Jose Oretga y Gasset — THE MODERN THEME. Introduction by Jose Ferrater Mora TB/1038
J. H. Parry — THE ESTABLISHMENT OF THE EUROPEAN HEGEMONY: 1415-1715: *Trade and Exploration in the Age of the Renaissance* TB/1045
Robert Payne — HUBRIS: *A Study of Pride.* Foreword by Herbert Read TB/1031
Samuel Pepys — THE DIARY OF SAMUEL PEPYS: Selections, edited by O. F. Morshead; illustrated by Ernest H. Shepard TB/1007
Paul E. Pfeutze — SELF, SOCIETY, EXISTENCE: *Human Nature and Dialogue in the Thought of George Herbert Mead and Martin Buber* TB/1059
Georges Poulet — STUDIES IN HUMAN TIME: *Montaigne, Molière, Baudelaire, Proust, et al.* TB/1004
George E. Probst, *Ed.* — THE HAPPY REPUBLIC: *A Reader in Tocqueville's America* TB/1060
Priscilla Robertson — REVOLUTIONS OF 1848: *A Social History* TB/1025
Ferdinand Schevill — THE MEDICI. Illustrated TB/1010
Bruno Snell — THE DISCOVERY OF THE MIND: *The Greek Origins of European Thought* TB/1018
C. P. Snow — TIME OF HOPE. A novel TB/1040
Perrin Stryker — THE CHARACTER OF THE EXECUTIVE: *Eleven Studies in Managerial Qualities* TB/1041
Percy Sykes — A HISTORY OF EXPLORATION. Introduction by John K. Wright TB/1046
Dorothy Van Ghent — THE ENGLISH NOVEL: *Form and Function* TB/1050
W. H. Walsh — PHILOSOPHY OF HISTORY: *An Introduction* TB/1020
W. Lloyd Warner — SOCIAL CLASS IN AMERICA: *The Evaluation of Status* TB/1013
Alfred N. Whitehead — PROCESS AND REALITY: *An Essay in Cosmology* TB/1033
Louis B. Wright — CULTURE ON THE MOVING FRONTIER TB/1053

HARPER TORCHBOOKS / The Science Library

Angus d'A. Bellairs — REPTILES: *Life History, Evolution, and Structure.* Illustrated TB/520
L. von Bertalanffy — PROBLEMS OF LIFE: *An Evaluation of Modern Biological and Scientific Thought* TB/521
David Bohm — CAUSALITY AND CHANCE IN MODERN PHYSICS. Foreword by Louis de Broglie TB/536
R. B. Braithwaite — SCIENTIFIC EXPLANATION TB/515
P. W. Bridgman — THE NATURE OF THERMODYNAMICS TB/537
Louis de Broglie — PHYSICS AND MICROPHYSICS. Foreword by Albert Einstein TB/514
J. Bronowski — SCIENCE AND HUMAN VALUES TB/505
A. J. Cain — ANIMAL SPECIES AND THEIR EVOLUTION. Illustrated TB/519
R. E. Coker — THIS GREAT AND WIDE SEA: *An Introduction to Oceanography and Màrine Biology*. Illustrated TB/551
T. G. Cowling — MOLECULES IN MOTION: *An Introduction to the Kinetic Theory of Gases.* Illustrated TB/516
A. C. Crombie, *Ed.* — TURNING POINTS IN PHYSICS TB/535
W. C. Dampier, *Ed.* — READINGS IN THE LITERATURE OF SCIENCE. Illustrated TB/512
H. Davenport — THE HIGHER ARITHMETIC: *An Introduction to the Theory of Numbers* TB/526
W. H. Dowdeswell — ANIMAL ECOLOGY. Illustrated TB/543
W. H. Dowdeswell — THE MECHANISM OF EVOLUTION TB/527
C. V. Durell — READABLE RELATIVITY TB/530
Arthur Eddington — SPACE, TIME AND GRAVITATION: *An Outline of the General Relativity Theory* TB/510
Alexander Findlay — CHEMISTRY IN THE SERVICE OF MAN. Illustrated TB/524
H. G. Forder — GEOMETRY: *An Introduction.* Illus. TB/548
Gottlob Frege — THE FOUNDATIONS OF ARITHMETIC TB/534
R. W. Gerard — UNRESTING CELLS. Illustrated TB/541
Werner Heisenberg — PHYSICS AND PHILOSOPHY: *The Revolution in Modern Science* TB/549
C. Judson Herrick — THE EVOLUTION OF HUMAN NATURE TB/545
Max Jammer — CONCEPTS OF SPACE TB/533
Max Jammer — CONCEPTS OF FORCE TB/550
S. Korner — THE PHILOSOPHY OF MATHEMATICS: *An Introduction* TB/547
David Lack — DARWIN'S FINCHES: *The General Biological Theory of Evolution.* Illustrated TB/544

D. E. Littlewood — THE SKELETON KEY OF MATHEMATICS: *A Simple Account of Complex Algebraic Theories* TB/525

J. E. Morton — MOLLUSCS: *An Introduction to Their Form and Function.* Illustrated TB/529

O. Neugebauer — THE EXACT SCIENCES IN ANTIQUITY TB/552

J. R. Partington — A SHORT HISTORY OF CHEMISTRY. Illustrated TB/522

H. T. Pledge — SCIENCE SINCE 1500: *A Short History of Mathematics, Physics, Chemistry, and Biology.* Illustrated TB/506

John Read — A DIRECT ENTRY TO ORGANIC CHEMISTRY. Illustrated TB/523

O. W. Richards — THE SOCIAL INSECTS. Illustrated TB/542

George Sarton — ANCIENT SCIENCE AND MODERN CIVILIZATION TB/501

Paul A. Schilpp, *Ed.* — ALBERT EINSTEIN: *Philosopher-Scientist. Vol. I,* TB/502; *Vol. II,* TB/503

P. M. Sheppard — NATURAL SELECTION AND HEREDITY. Illustrated TB/528

Edmund W. Sinnott — CELL AND PSYCHE: *The Biology of Purpose* TB/546

L. S. Stebbing — A MODERN INTRODUCTION TO LOGIC TB/538

O. G. Sutton — MATHEMATICS IN ACTION. Foreword by James R. Newman. Illustrated TB/518

Stephen Toulmin — THE PHILOSOPHY OF SCIENCE: *An Introduction* TB/513

A. G. Van Melsen — FROM ATOMOS TO ATOM: *The History of the Concept* Atom TB/517

Friedrich Waismann — INTRODUCTION TO MATHEMATICAL THINKING. Foreword by Karl Menger TB/511

W. H. Watson — ON UNDERSTANDING PHYSICS: *An Analysis of the Philosophy of Physics.* Introduction by Ernest Nagel TB/507

G. J. Whitrow — THE STRUCTURE AND EVOLUTION OF THE UNIVERSE: *An Introduction to Cosmology.* Illustrated TB/504

Edmund Whittaker — HISTORY OF THE THEORIES OF AETHER AND ELECTRICITY. *Vol. I, The Classical Theories,* TB/531; *Vol. II, The Modern Theories,* TB/532

A. Wolf — A HISTORY OF SCIENCE, TECHNOLOGY, AND PHILOSOPHY IN THE SIXTEENTH AND SEVENTEENTH CENTURIES. Illustrated. *Vol. I,* TB/508; *Vol. II,* TB/509

A. Wolf — A HISTORY OF SCIENCE, TECHNOLOGY, AND PHILOSOPHY IN THE EIGHTEENTH CENTURY. *Vol. I,* TB/539; *Vol. II,* TB/540

HARPER TORCHBOOKS / The Cloister Library

Tor Andrae — MOHAMMED: *The Man and His Faith* TB/62

Augustine/Przywara — AN AUGUSTINE SYNTHESIS TB/35

Roland H. Bainton — THE TRAVAIL OF RELIGIOUS LIBERTY TB/30

C. K. Barrett, *Ed.* — THE NEW TESTAMENT BACKGROUND: *Selected Documents* TB/86

Karl Barth — DOGMATICS IN OUTLINE TB/56

Karl Barth — THE WORD OF GOD AND THE WORD OF MAN TB/13

Nicolas Berdyaev — THE BEGINNING AND THE END TB/14

Nicolas Berdyaev — THE DESTINY OF MAN TB/61

Anton T. Boisen — THE EXPLORATION OF THE INNER WORLD: *A Study of Mental Disorder and Religious Experience* TB/87

J. H. Breasted — DEVELOPMENT OF RELIGION AND THOUGHT IN ANCIENT EGYPT TB/57

Martin Buber — ECLIPSE OF GOD: *Studies in the Relation Between Religion and Philosophy* TB/12

Martin Buber — MOSES: *The Revelation and the Covenant* TB/27

Martin Buber — THE PROPHETIC FAITH TB/73

Martin Buber — TWO TYPES OF FAITH: *The interpenetration of Judaism and Christianity* TB/75

R. Bultmann, et al. — KERYGMA AND MYTH: *A Theological Debate.* Ed. by H. W. Bartsch TB/80

Jacob Burckhardt — THE CIVILIZATION OF THE RENAISSANCE IN ITALY. Illustrated Edition. Introduction by Benjamin Nelson and Charles Trinkaus. *Vol. I,* TB/40; *Vol. II,* TB/41

Emile Cailliet — PASCAL: *The Emergence of Genius* TB/82

Edward Conze — BUDDHISM: *Its Essence and Development* TB/58

Frederick Copleston — MEDIEVAL PHILOSOPHY TB/76

F. M. Cornford — FROM RELIGION TO PHILOSOPHY: *A Study in the Origins of Western Speculation* TB/20

G. G. Coulton — MEDIEVAL FAITH AND SYMBOLISM [Part I of "Art and the Reformation"]. Illustrated TB/25

G. G. Coulton — THE FATE OF MEDIEVAL ART IN THE RENAISSANCE AND REFORMATION [Part II of "Art and the Reformation"]. Illustrated TB/26

H. G. Creel — CONFUCIUS AND THE CHINESE WAY TB/63

Adolf Deissmann — PAUL: *A Study in Social and Religious History* TB/15

C. H. Dodd — THE AUTHORITY OF THE BIBLE TB/43

Johannes Eckhart — MEISTER ECKHART: A Modern Translation TB/8

Mircea Eliade — COSMOS AND HISTORY: *The Myth of the Eternal Return* TB/50

Mircea Eliade — THE SACRED AND THE PROFANE: *The Significance of Religious Myth, Symbolism, and Ritual Within Life and Culture* TB/81

Morton S. Enslin — CHRISTIAN BEGINNINGS TB/5

Morton S. Enslin — THE LITERATURE OF THE CHRISTIAN MOVEMENT TB/6

G. P. Fedotov — THE RUSSIAN RELIGIOUS MIND: *Kievan Christianity, the 10th to the 13th Centuries* TB/70

Ludwig Feuerbach — THE ESSENCE OF CHRISTIANITY. Introduction by Karl Barth TB/11

Harry E. Fosdick — A GUIDE TO UNDERSTANDING THE BIBLE TB/2

Henri Frankfort — ANCIENT EGYPTIAN RELIGION: *An Interpretation* TB/77

Sigmund Freud — ON CREATIVITY AND THE UNCONSCIOUS: *Papers on the Psychology of Art, Literature, Love, Religion.* Edited by Benjamin Nelson TB/45

Maurice Friedman — MARTIN BUBER: *The Life of Dialogue* TB/64

O. B. Frothingham — TRANSCENDENTALISM IN NEW ENGLAND: *A History* TB/59

Edward Gibbon — THE TRIUMPH OF CHRISTENDOM IN THE ROMAN EMPIRE [J. B. Bury Edition, illus., Chapters 15-20 of "The Decline and Fall"] TB/46

C. C. Gillispie — GENESIS AND GEOLOGY: *A Study in the Relations of Scientific Thought, Natural Theology, and Social Opinion in Great Britain, 1790-1850* TB/51

Maurice Goguel — JESUS AND THE ORIGINS OF CHRISTIANITY I: *Prolegomena to the Life of Jesus* TB/65

Maurice Goguel — JESUS AND THE ORIGINS OF CHRISTIANITY II: *The Life of Jesus* TB/66

Edgar J. Goodspeed — A LIFE OF JESUS TB/1

H. J. C. Grierson — CROSS-CURRENTS IN 17TH CENTURY ENGLISH LITERATURE: *The World, the Flesh, the Spirit* TB/47

William Haller — THE RISE OF PURITANISM TB/22

Adolf Harnack — WHAT IS CHRISTIANITY? Introduction by Rudolf Bultmann TB/17

R. K. Harrison — THE DEAD SEA SCROLLS: *An Introduction* TB/84

Edwin Hatch — THE INFLUENCE OF GREEK IDEAS ON CHRISTIANITY TB/18

Friedrich Hegel — ON CHRISTIANITY: *Early Theological Writings* TB/79

Karl Heim — CHRISTIAN FAITH AND NATURAL SCIENCE TB/16

F. H. Heinemann — EXISTENTIALISM AND THE MODERN PREDICAMENT TB/28

S. R. Hopper, *Ed.* — SPIRITUAL PROBLEMS IN CONTEMPORARY LITERATURE TB/21

Johan Huizinga — ERASMUS AND THE AGE OF REFORMATION. Illustrated TB/19

Aldous Huxley — THE DEVILS OF LOUDUN: *A Study in the Psychology of Power Politics and Mystical Religion in the France of Cardinal Richelieu* TB/60

Flavius Josephus — THE GREAT ROMAN-JEWISH WAR, with *The Life of Josephus.* TB/74

Immanuel Kant — RELIGION WITHIN THE LIMITS OF REASON ALONE. TB/67

Soren Kierkegaard — EDIFYING DISCOURSES: A Selection TB/32

Soren Kierkegaard — THE JOURNALS OF KIERKEGAARD: A Selection. Edited by A. Dru TB/52

Soren Kierkegaard — PURITY OF HEART TB/4

Soren Kierkegaard — THE POINT OF VIEW FOR MY WORK AS AN AUTHOR: *A Report to History* TB/88

Alexandre Koyré — FROM THE CLOSED WORLD TO THE INFINITE UNIVERSE TB/31

Walter Lowrie — KIERKEGAARD. *Vol I,* TB/89; *Vol. II,* TB/90

Emile Mâle — THE GOTHIC IMAGE: *Religious Art in France of the 13th Century.* Illustrated TB/44

T. J. Meek — HEBREW ORIGINS TB/69

H. Richard Niebuhr — CHRIST AND CULTURE TB/3

H. Richard Niebuhr — THE KINGDOM OF GOD IN AMERICA TB/49

Martin P. Nilsson — GREEK FOLK RELIGION TB/78

H. J. Rose — RELIGION IN GREECE AND ROME TB/55

Josiah Royce — THE RELIGIOUS ASPECT OF PHILOSOPHY: *A Critique of the Bases of Conduct and Faith* TB/29

George Santayana — INTERPRETATIONS OF POETRY AND RELIGION TB/9

George Santayana — WINDS OF DOCTRINE *and* PLATONISM AND THE SPIRITUAL LIFE TB/24

F. Schleiermacher — ON RELIGION: *Speeches to Its Cultured Despisers* TB/36

H. O. Taylor — THE EMERGENCE OF CHRISTIAN CULTURE IN THE WEST: *The Classical Heritage of the Middle Ages* TB/48

P. Teilhard de Chardin — THE PHENOMENON OF MAN TB/83

D. W. Thomas, *Ed.* — DOCUMENTS FROM OLD TESTAMENT TIMES TB/85

Paul Tillich — DYNAMICS OF FAITH TB/42

Ernst Troeltsch — THE SOCIAL TEACHING OF THE CHRISTIAN CHURCHES. Introduction by H. Richard Niebuhr. *Vol. I,* TB/71; *Vol. II,* TB/72

E. B. Tylor — THE ORIGINS OF CULTURE [Part I of "Primitive Culture"]. Introduction by Paul Radin TB/33

E. B. Tylor — RELIGION IN PRIMITIVE CULTURE [Part II of "Primitive Culture"]. Introduction by Paul Radin TB/34

Evelyn Underhill — WORSHIP TB/10

Johannes Weiss — EARLIEST CHRISTIANITY: *A History of the Period* A.D. *30-150.* Introduction by F. C. Grant. *Vol. I,* TB/53; *Vol. II,* TB/54

Wilhelm Windelband — A HISTORY OF PHILOSOPHY I: *Greek, Roman, Medieval* TB/38

Wilhelm Windelband — A HISTORY OF PHILOSOPHY II: *Renaissance, Enlightenment, Modern* TB/39